SpringerBriefs in Electrical and Computer Engineering

T0224082

For further volumes:
http://www.springer.com/series/10059

Ilya Gertsbakh · Yoseph Shpungin

Network Reliability
and Resilience

Springer

Prof. Dr. Ilya Gertsbakh
Department of Mathematics
Ben Gurion University
84105 Beer-Sheva
Israel
e-mail: elyager@bezeqint.net

Dr. Yoseph Shpungin
Department of Software Engineering
Shamoon College of Engineering
84100 Beer-Sheva
Israel
e-mail: yosefs@sce.ac.il

ISSN 2191-8112   e-ISSN 2191-8120
ISBN 978-3-642-22373-0   e-ISBN 978-3-642-22374-7
DOI 10.1007/978-3-642-22374-7
Springer Heidelberg Dordrecht London New York

Cover design: eStudio Calamar, Berlin/Figueres

Printed on acid-free paper

Springer is part of Springer Science+Business Media (www.springer.com)

Preface

This book is a result of our ongoing research on network reliability. Started in 1991 [1], it was first summarized in *"Models of Network Reliability: Analysis, Combinatorics and Monte Carlo"* [2]. Recently, we have widened our approach to include networks with many (more than two) states. In [2] we considered only networks with two states: *UP* and *DOWN*, mainly for the situation where *DOWN* meant loss of terminal connectivity. This was a relatively narrow approach and it becomes much more comprehensive when we assume that a network can have many states that reflect its degradation in the process of nodes or links failures. Such a comprehensive approach enables to describe probabilistically the process of network gradual disintegration into isolated clusters, which starts when all terminals are connected to each other and ends with partial or total isolation of all terminals. Another option that such an approach affords is to follow the size of the network's largest connected component when network nodes are subject to random "attack".

The main formal tools for our investigation are the so-called multidimensional D-spectrum and the marginal D-spectra. D-spectrum is an object of combinatorial nature and is completely determined by the system structure function. This allows to develop efficient Monte Carlo procedures for approximating system D-spectra and serves as a basis for numerical analysis and reliability calculations in the area of network reliability and resilience.

The exposition is as follows. The first chapter is devoted to the theory. It starts with a brief summary of a traditional material on reliability of monotone binary systems and their applications to networks (Sects. 1.1 and 1.2), with the difference being that we extend the definition of the binary system to the case of more than two states.

Section 1.3 contains the definition of the D-spectrum and the marginal D-spectra. In simple words, D-spectrum is a multidimensional discrete probability distribution, whose rth coordinate is the discrete distribution of the number of the component whose failure causes the transition of the network from state $J + 1$ into state J. In binary situation with only one such transition, the D-spectrum numerically coincides with so-called *signature* introduced by Samaniego in [3] and is equivalent to so-called *Internal Distribution* suggested by Lomonosov in [1].

Sections 1.3–1.5 contain extensions of the D-spectra to recurrent networks and series and parallel connection and network-type systems, as well to the case when network components may have several states. The latter we call networks with colored links.

One of the most important issues in network reliability is the network design aimed at improving its reliability parameters. The central role here belongs to the reliability gradient function which shows the increase of system reliability as a function of component reliability increase. In the case of independent and equally reliable components the gradient coincides with so-called *Birnbaum Importance Measure* (BIM) [4].

Even for small networks we meet the situation when the analytic formula for network reliability is not available and the straightforward computation of the gradient function becomes practically impossible.

It turns out, however, that the BIM of a component can be estimated via a network combinatorial parameter, so-called BIM-spectrum, which is closely related to the network D-spectrum (Sect. 1.6). An extension of the BIM-spectrum allows to obtain a combinatorial formula for another important index, so-called *Joint Reliability Index* [5], which is the second mixed derivative of network reliability function.

In general case of arbitrary component reliabilities the calculation of gradient function becomes more involved but nevertheless can be carried out using another combinatorial characteristic of the network, so-called *border states*. In words, the border state is a network *DOWN* state which has a unit Manhattan distance from the network *UP* state. The material related to network gradient function is presented in Sect. 1.7.

We believe that a central issue in reliability engineering is the ability to efficiently calculate the numerical values of the theoretically derived reliability indices. Since the exact computations are as a rule NP-complete, the Monte Carlo approximations remain our main computation tool. Section 1.8 describes in a non formal way three main Monte Carlo procedures: estimation of network connectivity, estimation of the D-spectra and component BIMs, and the estimation of the gradient function. A quite natural question is the accuracy which can be provided by a limited, say $M = 10^5$ to 10^7 Monte Carlo replications, within reasonable time limits. We demonstrate that in typical cases, such as approximating the reliability of a five-dimensional cubic network with 32 nodes and 80 edges, $M = 10^6$ replications guarantees an absolute error not exceeding 7×10^{-4}, which is a satisfactory accuracy for engineering calculations.

Chapter 2 is devoted to applications. In Sect. 2.1 we compare the behavior of networks under random attack on their nodes. We present an example in which a regular network is more resilient than the scale-free network with the same number of nodes and links.

Section 2.2 describes various approaches to network reliability design based on improving their reliability by means of reinforcing several nodes or several links combined with deletion of the least important components.

Section 2.3 deals with an example of an optimal pre-disaster management of a transportation network. This is implemented by the "best" choice of the subset of links which are reinforced to provide given level of terminal connectivity, subject to budgetary constraint.

Finally, Sect. 2.4 is an example of a detailed probabilistic follow-up of the network disintegration into isolated clusters when the links fail in a random order.

We hope that this work will be of interest to reliability researchers involved in network design and study, and to reliability engineers interested in applications of the theory to practical calculations of network reliability parameters.

References

[1] Elperin, T., Gertsbakh, I., and M. Lomonosov. 1991. Estimation of network reliability using graph evolution models. *IEEE Transactions on Reliability* R-40: 572–581.
[2] Gertsbakh, Ilya, and Yoseph, Shpungin. 2009. *Models of network reliability: analysis, combinatorics, and Monte Carlo.* Boca Raton: CRC Press.
[3] Samaniego, F.J. 1985. On closure of the IFR under formation of coherent systems. *IEEE Transactions on Reliability* 34: 69–72.
[4] Birnbaum, Z.W. 1969. On the importance of different components in multi-component system. In *Multivariate analysis-II*, ed. P.R. Krishnaiah, 581–592. New York: Academic Press.
[5] Jong, S.H., and Chang H.L. 1993. Joint reliability importance of two edges in undirected network. *IEEE Transactions on Reliability* 42(1): 17–23.

May 2011   Ilya Gertsbakh
                                                                           Yoseph Shpungin

Contents

Notation and Abbreviations

i.i.d.	Independent identically distributed
i.r.v.	Independent random variables
r.v.	Random variable
c.d.f.	Cumulative distribution function (CDF)
d.f.	Density function
τ, X, Y, Z	Random variables
up, down	States of a binary component
UP, DOWN	States of a binary system
$X \sim U(0,1)$	r.v. X is uniformly distributed on [0,1]
$E[X]$	Mathematical expectation (mean value) of r.v. X
$\text{Var}[X]$	Variance of r.v. X
σ	Square root of variance, $\sigma = \sqrt{\text{Var}[X]}$, also termed standard deviation
σ_X	Standard deviation of r.v. X
$X \sim Exp(\lambda)$	r.v. X has an exponential distribution with parameter $\lambda: P(X \leq t) = 1 - \exp(-\lambda t), t \geq 0$
$r.e.[X]$	Relative error of r.v. X. Defined as $\sigma_X / E[X]$, for nonnegative r.v.s only
$X \sim \Diamond(\mu, \sigma)$	r.v. X has mean value μ and st.dev σ
BIM	Birnbaum importance measure
BIM_j	Birnbaum importance measure of component j
JRI	Joint reliability index
x_i	Indicator variable of binary component i
$\mathbf{x} = (x_1, x_2, ..., x_n)$	System component state vector
$\varphi(\cdot)$	System structure function

X_i	Random indicator variable of binary component i
$\mathbf{f}^{(s)} = (f_1^{(s)}, f_2^{(s)}, \ldots, f_n^{(s)})$	The sth marginal D-spectrum
$F^{(s)}(x) = \sum_{i=1}^{x} f_i^{(s)}, \quad x = 1, 2, \ldots, n$	The sth cumulative marginal D-spectrum
$R = \Psi(p_1, p_2, \ldots, p_k)$	System reliability as a function of components reliability
$\nabla R = (\partial R / \partial p_1, \ldots, \partial R / \partial p_n)$	System gradient vector
J	Network state. $J = 0$ denotes network DOWN state
n, k	Number of components in the network (nodes or links)
Y	The random number of components whose failure causes the network to change its state, for example, from UP to DOWN
$P_N(J; p)$	Probability that the network N is in state J
$X_{i:n}$	The ith order statistic from the random sample of $\{X_1, \ldots, X_n\}$
$res_{pr}(\mathbf{N}; \beta)$	Probabilistic resilience of network N
$\eta = res_{pr}(\mathbf{N}; \beta)/n$	Probabilistic resilience rate; n is the number of nodes or links in the network
$\mathbf{N} = (V, E, T)$	Network with vertex (node) set V, edge (link) set E and terminal set T
$\bar{d}$	Average node degree

Chapter 1
Theory

Abstract Sections 1 and 2 present brief summary of a standard material on reliability of monotone binary systems and their applications to networks. The definition
of the binary system is extended to the case of more than two states. Section 3 contains the definition of the D-spectrum and the marginal D-spectra. D-spectrum is a
multidimensional discrete probability distribution, whose rth coordinate is a discrete
distribution of the number of the component whose failure causes the transition of
the network from state $j + 1$, into state j computed under assumption that components fail in random order. D-spectrum is a combinatorial characteristic of the
system which in particular case of two-state system with i.i.d. components coincides
with Samaniego's signature. Network probabilistic resilience presented in Section 2
is the $(1 - \beta)$-quantile of the cumulative marginal D-spectrum. Sections 3, 4 show
how to compute D-spectra for recurrent networks and series-parallel connection of
network-type systems. Section 5 considers networks with multi-state edges which
we call networks with colored links. Section 6 deals with Birnbaum Importance
Measure (BIM) of network components for networks with identical and independent components. It is shown that the BIM of component j can be estimated via a
network combinatorial parameter, so-called BIM-spectrum, which is closely related
to the network D-spectrum. An extension of the BIM-spectrum allows to obtain a
combinatorial formula for another important index, so-called *Joint Reliability Index*.
Section 7 discusses reliability gradient function which allows to compute the increase
of system reliability as a function of component reliability increase. In general case of
arbitrary component reliabilities the calculation of gradient function becomes more
involved but nevertheless can be carried out by using another combinatorial characteristic of the network, so-called *border states*. Border state is a network *DOWN*
state which has a unit Manhattan distance from the network *UP* state. As a rule,
most of computations aimed at network reliability estimation are NP-complete, and
the Monte Carlo approximations remain our main computation tool. Section 1.8
describes in a non formal way three principal Monte Carlo procedures: estimation
of network connectivity, estimation of the D-spectra and component BIM's, and
the estimation of the gradient function. It discusses also the question of accuracy

I. Gertsbakh and Y. Shpungin, *Network Reliability and Resilience*,
SpringerBriefs in Electrical and Computer Engineering,
DOI: 10.1007/978-3-642-22374-7_1, © Ilya Gertsbakh 2011

provided by a limited number of Monte Carlo replications which guarantee a numerical result with an error small enough for engineering calculations. Finally, Sect. 9 discusses the construction of D-spectra for the case when a single component failure (e.g. node failure) leads the network to change its state J to state $J - k$, $k \geq 2$.

Keywords Multi-state network · Multi-state D-spectra · Component importance measure · Gradient function · Border states · Monte Carlo accuracy

1.1 Graphs, Networks, Terminals, Structure Function

We meet networks every day and everywhere in our life. For formal study of network properties we must operate with abstract *models* of networks. In further, our principal network model will be a triple $\mathbf{N} = (V, E, T)$, where V is the *vertex* or *node* set, $|V| = m$, E is the *edge* or *link* set, $|E| = n$, and T is a set of special nodes called *terminals*, $T \subseteq V$, $|T| = h$.

In simple words, a network is a collection of circles (nodes) and links, i.e. line segments connecting the nodes. Terminals are marked as bold circles, like in Fig. 1.1.

In most situations we will be dealing with networks having several *states*, according to the presence or absence of connection between the terminals. We say that two terminals a and b are *connected* if there is a *path* of links connecting them. For example, the path $(a, d) \leftrightarrow (d, e) \leftrightarrow (e, b)$ connects terminals a and b.

If all terminals are connected to each other, we say that the network is *T-connected*. A subset $V_1 \subset V$ is called an *isolated component* of $\mathbf{N}$ if all nodes in V_1 are connected to each other and there are no edges of type $e = (a, b)$, where $a \in V_1$ and $b \in V - V_1$. An isolated node is considered as isolated component. An isolated component of $\mathbf{N}$ is called a *cluster* if it contains at least one terminal node. A single terminal node is considered as a cluster.

The network on Fig. 1.1 has only one cluster. If the links $1 = (a, d)$, $7 = (f, e)$, $8 = (f, b)$, $4 = (g, c)$ and $10 = (g, b)$ will be erased, then there will be two (isolated) clusters in $\mathbf{N}$: one cluster will contain terminal a and another—terminals b and c. Suppose that links $1 = (a, d)$, $7 = (f, e)$, $9 = (e, b)$ and $3 = (e, c)$ are erased. Then the network falls apart into two isolated components, one of which is a cluster.

Our exposition will be centered around network behavior when its elements (nodes and/or links) fail. We will deal mainly with so-called *binary* elements which can be in two states *up* and *down* denoted by 1 and 0, respectively. When speaking about links, link i failure means that this link is erased, i.e. it does not exist. The state of link i, $i = 1, \ldots, n$ is denoted by binary variable x_i. If $x_i = 1$, link i is *up*; if $x_i = 0$, link i is *down*. x_i is often called link indicator variable. In some models, the elements subject to failure are network *nodes* (vertices). If the indicator variable of node j is $y_j = 0$, i.e. node j is *down*, it means that all links incident to node j are *erased*, but the node itself remain intact. So, for example, if node f on Fig. 1.1 is *down*, all four links $6 = (a, f)$, $7 = (e, f)$, $8 = (b, f)$ and $5 = (g, f)$ are erased. By agreement, the terminals *do not fail*.

Fig. 1.1 Network with 11 links, seven nodes and three terminals (*bold*)

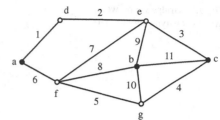

Fig. 1.2 Representation of a series system (**a**) and parallel connection of series systems (**b**)

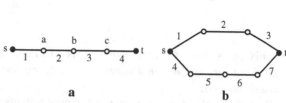

The next step will be introducing a compact description of all network element states. We will use vector notation $\mathbf{x} = (x_1, x_2, x_3, \ldots, x_n)$. For example, if all odd numbered links are down, and all even numbered links are *up*, network state vector for Fig. 1.1 will be $\mathbf{x} = (0, 1, 0, 1, 0, 1, 0, 1, 0, 1, 0)$.

It will be assumed that the network can be in several states denoted by integers $J = 0, 1, 2, \ldots, L$. The dependence of network's state on the state of its elements (nodes or links) will be determined by means of the so-called *structure function* $\varphi(\mathbf{x})$. Let us consider several examples of typical structure functions.

Example 1.1.1 (*Simplest s–t network*, Fig. 1.2a)
This network has two terminals s and t and four links $1 = (s, a), 2 = (a, b), 3 = (b, c), 4 = (c, t)$ which are subject to failure. By definition, the network has two states, *UP* ($J = 1$), if all links are *up* or *DOWN* ($J = 0$), otherwise. Formally,

$$\varphi(\mathbf{x}) = \prod_{i=1}^{4} x_i = \min_{1 \leq i \leq 4} x_i. \tag{1.1.1}$$

In reliability theory, this network is called *series system*.

Example 1.1.2 (*Two s–t networks in parallel*, Fig. 1.2b)
This network has also two terminals s and t and two "paths" of links in parallel. By definition, it is *UP* if and only if there is a connection between s and t. Elements subject to failure are links. Formally,

$$\varphi(\mathbf{x}) = 1 - \left(1 - \prod_{i=1}^{3} x_i\right)\left(1 - \prod_{i=4}^{7} x_i\right) = \max(x_1 x_2 x_3, x_4 x_5 x_6 x_7). \tag{1.1.2}$$

In reliability theory, this network is called a *parallel* connection of two series systems.

It is desirable to have a systematic way of constructing a formula for the structure function $\varphi(\cdot)$. It is relatively easy to do this for a binary case when the network

Fig. 1.3 Bridge *s–t* network
(a) and star-type network (**b**)

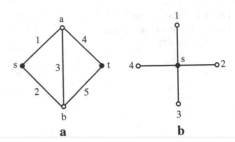

has only two states—*UP* and *DOWN*. This will be done by introducing the notions of *minimal cuts* and *minimal paths*. Before doing this let us impose some natural demands on $\varphi(\cdot)$.

We will use the notation $\mathbf{x} < \mathbf{y}$ if the components of $\mathbf{x}$ are less or equal to the components of $\mathbf{y}$, but there is at least one component j such that $x_j < y_j$.

Definition 1.1.1 (*Monotone system*) A network with structure function $\varphi(\cdot)$ is called *monotone* if it has the following properties:

(i) $\varphi(0, 0, \ldots, 0) = 0$, $\varphi(1, 1, \ldots, 1) = L > 0$;
(ii) $\mathbf{x} < \mathbf{y} \Rightarrow \varphi(\mathbf{x}) \leq \varphi(\mathbf{y})$.

In words: the network is *DOWN* ($J = 0$) if all its elements are *down*; if all elements are *up*, the network state is $L > 0$; the state of the network can not become worse if any of its elements change its state from *down* to *up*.

Let us assume that the network is binary, i.e. it can be in two states *UP* ($J = 1$) or *DOWN* ($J = 0$). For this situation we introduce important notions of *path set, path vector, cut set* and *cut vector*.

Definition 1.1.2 (*Cut vector, cut set, path vector, path set*)
A state vector $\mathbf{x}$ is called a *cut vector* if $\varphi(\mathbf{x}) = 0$. The set $C_t(\mathbf{x}) = \{i : x_i = 0\}$ is called a *cut set*. If, in addition, for any $\mathbf{y} > \mathbf{x}$, $\varphi(\mathbf{y}) = 1$, then the corresponding cut set is called *minimal cut set*, or simply *minimal cut*. A state vector $\mathbf{x}$ is called a *path vector* if $\varphi(\mathbf{x}) = 1$. The set $P_t(\mathbf{x}) = \{i : x_i = 1\}$ is called a *path set*. If, in addition, for any $\mathbf{y} < \mathbf{x}$, $\varphi(\mathbf{y}) = 0$, then the corresponding path set is called *minimal path set*, or simply *minimal path*.

Let us illustrate these notions by an example of a bridge-type network shown on Fig. 1.3a.

Bridge network has two terminals, four nodes and five links. Links are subject to failures. The network is *UP* if and only if the connection between s and t does exist. $\mathbf{x} = (1, 1, 1, 0, 1)$ is a path vector and $\{1, 2, 3, 5\}$ is the corresponding path set, but not a minimal path set since it can be reduced to $\{1, 3, 5\}$ and remains a path set. The latter is a minimal path. Similarly, $\mathbf{x} = (1, 1, 1, 0, 0)$ is a cut vector, but $\{1, 2, 3\}$ is not a minimal cut. The set $\{1, 2\}$ is, on the contrary, a minimal cut set.

Now we are ready to represent the central result of this section.

Theorem 1.1.1 (Structure function representation)
Suppose that the network has a binary structure function. Let $P_{t_1}, P_{t_2}, \ldots, P_{t_s}$ be the minimal path sets of the network and $C_{t_1}, C_{t_2}, \ldots, C_{t_k}$ be the minimal cut sets of the network. Then

$$\varphi(\mathbf{x}) = 1 - \prod_{j=1}^{s}\left(1 - \prod_{i \in P_{t_j}} x_i\right) = \prod_{j=1}^{k}\left(1 - \prod_{i \in C_{t_j}}(1 - x_i)\right). \qquad (1.1.3)$$

Proof Assume that there is at least one minimal path set, all elements of which are *up*, say P_{t_1}. Then $\prod_{i \in P_{t_1}} x_i = 1$ and this leads to $\varphi(\mathbf{x}) = 1$. Suppose now that the network is *UP*. Then there must be at least one minimal path having all its elements in the *up* state. This proves the first equality. The proof of the second equality is similar and we omit it.   □

For small networks, where enumeration of all minimal cut sets or all minimal path sets is a feasible problem, this theorem opens a way to network reliability calculations.

Let us conclude this section with several examples of defining the structure function $\varphi(\cdot)$.

a. *Flow in the network.* Consider again the bridge network shown on Fig. 1.3a. Suppose that each link has capacity 1 and network state is determined as the maximal flow which can be delivered from s to t. The elements of network subject to failure are the links. Obviously, $\varphi(1, 1, 1, 1, 1) = \varphi(1, 1, 0, 1, 1) = 2$. If one of the links 1, 2, 4 or 5 is *down*, the flow from s to t drops by one and $\varphi(\cdot) = 1$. If all links of one min cut set are *down*, then $\varphi(\cdot) = 0$, for example, $\varphi(0, 1, 0, 1, 0) = 0$.

b. *Three-terminal connectivity.* Suppose that in the bridge network, nodes s, t *and a* are terminals. Again the links are subject to failure. The function $\varphi(\cdot)$ is set to be 1 (*UP*) if and only if all terminals are connected to each other, i.e. there is only one cluster in the network. Otherwise, $\varphi(\cdot) = 0$. For example, $\varphi(1, 0, 0, 1, 0) = 1$ and $\varphi(0, 1, 0, 0, 1) = 0$. Indeed if only links 2 and 5 are *up*, terminal *a* becomes isolated and the network is *DOWN*, by definition.

c. *Several isolated components in the network.* Figure 1.3b shows a small star-type network with five nodes. Elements of the network subject to failure are the nodes. $\varphi(\cdot)$ is set to be equal to 5 minus the number of isolated components in the network. Node failure means that all edges incident to this node are erased. So, if all nodes are connected to each other into one component, $\varphi(\cdot)$ is 4. If a single node 1, 2, 3 or 4 fails, the number of isolated components increases by one. If node s fails, the networks falls apart into 5 isolated components and $\varphi(\cdot)$ becomes 0.

1.2 Network Reliability

1.2.1 Binary Networks with Independent Binary Components

Contrary to the previous section, let us now assume that the state of network component i (node or link) is described by a binary *random variable* X_i, defined as

$$P(X_i = 1) = p_i, \quad P(X_i = 0) = 1 - p_i = q_i, \tag{1.2.1}$$

where 1 and 0 correspond to *up* and *down* state, respectively.

It will be assumed that all components are mutually independent. This implies that the joint distribution of $X_1, X_2, \ldots, X_n$ is completely determined by component reliabilities $p_1, p_2, \ldots, p_n$.

Denote by $\mathbf{X} = (X_1, X_2, \ldots, X_n)$ the network state vector which now is a random vector. Correspondingly, the network structure function $\varphi(\mathbf{X})$ becomes a binary random variable: $\varphi(\mathbf{X}) = 1$ corresponds to network *UP* state and $\varphi(\mathbf{X}) = 0$—to network *DOWN* state.

Definition 1.2.1 (*Network reliability*)
Network reliability R_0 is the probability that the structure function equals 1:

$$R_0 = P(\varphi(\mathbf{X}) = 1). \tag{1.2.2}$$

Since $\varphi(\cdot)$ is binary, the last formula can be rewritten as

$$R_0 = E\left[\varphi(\mathbf{X})\right] = \Psi(p_1, p_2, \ldots, p_n). \tag{1.2.3}$$

Formula (1.2.3) is very useful since the operation of taking expectation $E[\cdot]$ considerably simplifies reliability calculations.

In the future we will need an important relationship called *pivotal decomposition*, see [1, 8]. Denote by $\Psi(p_1, p_2, \ldots, 1_j, \ldots, p_n)$ the probability of our network when its jth component is replaced by an absolutely reliable (permanently *up*), $p_j : = 1$. Similarly, $\Psi(p_1, p_2, \ldots, 0_j, \ldots, p_n)$ will be the network reliability when its jth component is permanently *down*, $p_j := 0$.

Theorem 1.2.1 (Pivotal decomposition)

$$R_0 = p_j \cdot \Psi(p_1, p_2, \ldots, 1_j, \ldots, p_n) + (1 - p_j) \cdot \Psi(p_1, p_2, \ldots, 0_j, \ldots, p_n). \tag{1.2.4}$$

The proof is elementary and based on the Law of Total Probability, see e.g. [8], Chap. 1.

Example 1.2.1 (*Reliability of series, parallel and bridge networks*)
The structure function for the network shown on Fig. 1.2a is $\varphi(\mathbf{X}) = X_1 X_2 X_3 X_4$. By (1.2.3),

$$R_0 = \prod_{i=1}^{4} p_i.$$

For the network on Fig. 1.2b,

$$\varphi(\mathbf{X}) = 1 - (1 - X_1 X_2 X_3)(1 - X_4 X_5 X_6 X_7).$$

Applying the $E[\cdot]$ operator and using the fact that the the expressions in the parentheses are independent, we obtain that

$$R_0 = 1 - (1 - p_1 p_2 p_3)(1 - p_4 p_5 p_6 p_7).$$

Unfortunately, not always we have series, parallel or series–parallel networks. The simplest example is the bridge network on Fig. 1.3a. There are four minimal path sets connecting s and t: $(1, 4),(2, 5),(1, 3, 5)$ and $(2, 3, 4)$. Using (1.1.3), we can represent system structure function as

$$\varphi(\mathbf{X}) = 1 - (1 - X_1 X_4)(1 - X_2 X_5)(1 - X_1 X_3 X_5)(1 - X_2 X_3 X_4).$$

Here the expressions in the brackets have common terms and therefore are *not* independent. Before we apply the $E[\cdot]$ operator, we must first open the brackets. Fortunately, algebraic operations are considerably simplified due to the fact that for binary variables $X_i^2 = X_i$. We present the final result for $p_i \equiv p$:

$$R_0(p) = 2p^2 + 2p^3 - 5p^4 + 2p^5.$$

This formula is called *reliability polynomial*. It should be noted that even for relatively small networks having 10–15 components, which are not of series–parallel type, the enumeration of all minimal path sets becomes a very difficult problem.

The situation considered in this section allows two physical interpretations. The first one is that the state of each component is determined by a random "lottery": component i is declared as being *up (down)* as a result of random lottery which produces the results with probability p_i and $1 - p_i = q_i$, respectively. Time coordinate is not present here, and one can imagine that all lotteries take place simultaneously or in some arbitrary sequence of time instants.

The second interpretation is related to a random process of operation—repair for network components. Assume that component i has random intervals $\xi_k^{(i)}$, $k = 1$, $2, 3, \ldots$, of operation (*up*-periods) alternating with random intervals of repair $\eta_k^{(i)}$, $k = 1, 2, 3, \ldots$, (*down*-periods). Suppose that $\xi_k^{(i)}$, $\eta_k^{(i)}$, $k = 1, 2, \ldots$, are independent and have mean values $\mu^{(i)}$ and $\nu^{(i)}$, respectively. The following quantity

$$A_v^{(i)} = \frac{\mu^{(i)}}{\mu^{(i)} + \nu^{(i)}} \tag{1.2.5}$$

is called *stationary availability* of component i. The physical meaning of $A_v^{(i)}$ is the following. Let $P^{(i)}(t)$ be the probability that component i is *up* at the time instant t. Then

$$\lim_{t \to \infty} P^{(i)}(t) = A_v^{(i)}. \tag{1.2.6}$$

Simply speaking, the stationary availability is the probability that component is *up* at some remote time moment, formally at $t \to \infty$. Another interpretation, see e.g. [1, 8] is the following. Denote by $V^{(i)}(T)$ the total amount of time on the interval $[0, T]$ during which component i was *up*. It can be proved that

$$A_v^{(i)} = \lim_{T \to \infty} E[V^{(i)}(T)]/T. \tag{1.2.7}$$

Now let us consider the expression $\varphi\left(A_v^{(1)}, A_v^{(2)}, \ldots, A_v^{(n)}\right) = A_v$. It is called system *stationary availability* and its physical meaning is the probability that the network is *UP* at some remote instant of time. Formal details of the proof of this statement can be found, e.g. in [1], Chap. 7.

1.2.2 Network Components with Independent Lifetimes

In this section, it will be assumed that network component i (node or link) has random lifetime τ_i with cumulative distribution function (CDF) $F_i(t) = P(\tau_i \leq t)$. $\{\tau_i\}$, $i = 1, 2, \ldots, n$, are independent positive continuous random variables, and at time $t = 0$ all components start their operation being in *up* state. After the component fails, it is not renewed and remains in state *down* forever.

It will be convenient to characterize the state of component i at time t by a binary random variable $X_i(t)$. $X_i(t) = 1$ if and only if $\tau_i > t$. If $\tau_i \leq t$, $X_i(t) = 0$. In other words, $X_i(t)$ is equal 1 as long as the component is *up*, and becomes 0 when the component goes *down*. Let $\mathbf{X}(t) = (X_1(t), X_2(t), \ldots, X_n(t))$ be the component vector state at time t.

Definition 1.2.2 (*Network lifetime*)
If the network has two states, *UP* and *DOWN*, its lifetime τ_N is the time until it enters the *DOWN* state:

$$\tau_N = \inf[t: \varphi(\mathbf{X}(t)) = 0]. \tag{1.2.8}$$

Let us define network reliability $R_0(t)$ as the probability that τ_N exceeds t:

$$R_0(t) = P(\tau_N > t).$$

R_0 is often called network survival function. In the previous section we have defined the "static" network reliability as $R_0 = E[\varphi(X_1, X_2, \ldots, X_n)] = \Psi(p_1, p_2, \ldots, p_n)$.

Theorem 1.2.1 (Network survival function)
Let $R_i(t) = P(\tau_i > t) = 1 - F_i(t)$. Then

$$R_0(t) = \Psi(R_1(t), R_2(t), \ldots, R_n(t)). \tag{1.2.9}$$

Proof By the definition of $R_i(t)$, $R_0(t)$ is the probability that the network is *UP* at the time instant t. Since the network is a monotone system, it was in state *UP* during the whole interval $[0, t]$, or $R_0(t) = P(\tau_N > t)$. □

This theorem says that if we have the expression of network reliability in the form of a function of component *up* probabilities p_i, just replace each p_i by $R_i(t)$ and we will have the network survival function.

The following example shows how to obtain the survival function for a series–parallel networks.

Example 1.2.2 (Minimum–maximum calculus)
Suppose we have a series-type s–t network with n links. Link i has lifetime τ_i. The network lifetime is, obviously, determined by the *minimum* of $\tau_1, \ldots, \tau_n$. The network survives time t if and only if all links do so. Therefore,

$$R_{\text{ser}}(t) = \prod_{i=1}^{n} (1 - F_i(t)) = \prod_{i=1}^{n} R_i(t).$$

Suppose now that terminals s,t are connected by n links in parallel. Obviously, this network lifetime $\tau_N = \max[\tau_1, \tau_2, \ldots, \tau_n]$. This network becomes *DOWN* if and only if all links are down. Therefore,

$$R_{\text{par}}(t) = 1 - \prod_{i=1}^{n} (1 - R_i(t)).$$

Now it is easy to handle a network which is a series or parallel combination of another series or parallel subnetworks. For example, consider network on Fig. 1.2b. Denote by τ_a the lifetime of the upper path of three links, and by τ_b—the lifetime of the lower path. Then $\tau_N = \max[\tau_a, \tau_b]$ and $R_0(t) = 1 - F_a(t)F_b(t)$. Here, obviously, $\tau_a = \min[\tau_1, \tau_2, \tau_3]$ and $\tau_b = \min[\tau_4, \ldots, \tau_7]$. Combining all together, we obtain that

$$R_0(t) = 1 - \left(1 - \prod_{i=1}^{3} (1 - F_i(t))\right)\left(1 - \prod_{i=4}^{7} (1 - F_i(t))\right).$$

In conclusion, let us present a very useful formula for computing the mean network lifetime. It can be proved that

$$E[\tau_N] = \int_0^\infty R_0(t)\, dt. \tag{1.2.10}$$

1.2.3 Concluding Remarks: Network Resilience

The theory which we presented in previous two sections may create an impression that we are able to solve easily all network reliability problems. This might be true if we could have an explicit analytic expression for the network structure function $\varphi(\cdot)$. A first warning that the situation is not so simple we got when we considered the bridge network. Fortunately, we could easily find out all minimal paths in this small network. Doing the same for larger network with, say 15 links, would be already a very difficult problem. In practice, the previously developed techniques are efficient only for a very narrow family of series–parallel s–t network-type systems.

Let us not forget another favorable fact which considerably simplified all calculations: in all examples in two previous sections we had a *binary* (1/0) structure function. In this case, we put into work the powerful operation of taking mathematical expectation which immediately provided the probability that the network is in *UP* state, see e.g. (1.2.3).

Real-life situations, even after formalization and simplification, are much more complex than the binary 0/1 case with known structure function. Consider, for example, the following network model. All network nodes are terminals, links have i.i.d. lifetimes with known CDF $G(t)$. The network state is determined as a function of the number of isolated clusters in it. For example, the state is defined as $J = 3$ if all terminals are connected (one cluster), $J = 2$ if there are exactly two clusters, $J = 1$ if the network disintegrates into three clusters, and finally $J = 0$ (*DOWN*) if the network falls apart into four or more clusters. We are interested in finding the probabilities $P_N(J; t)$ that at any given time instant t the network is in state J. Obviously, all our techniques developed so far are not sufficient to tackle this problem and more powerful tools are needed.

The classical network theory is mostly oriented on computer and communication networks. It approaches network reliability problems from somewhat different angle. If the "standard" reliability theory is interested in finding such principal quantities as probability of connectivity and/or network lifetime distribution function, network theory develops and studies probabilistic network *robustness statistics* in general and *probabilistic resilience*, in particular. *Wikipedia* gives the following definition:

"*Resilience* is the ability to provide and maintain an acceptable level of service in the face of faults and challenges to normal functioning operation."

Resilience is considered as a superset to *survivability* which is defined as "capability to fulfill network mission in the presence of attacking, failures and accidents of network components".

Both these definitions are too general to apply them in practice. We would like to present here a more formal version of network *probabilistic resilience* by citing the definitions from [3], Sect. 15.4.2, pp. 434–435.

In the case of random failures, the *disconnection probability* of a network **N** is defined as

$$P(\mathbf{N}; i) = P[\mathbf{N} \text{ is disconnected exactly after } i\text{th failure}].$$

Another important notion is presented in the following.

Definition 1.2.3 (*Probabilistic resilience* [3], p. 435)
Let $\mathbf{N}$ be a network with n components . The probabilistic resilience $\mathrm{res_{pr}}(\mathbf{N}; \beta)$ is the largest number of component failures such that $\mathbf{N}$ is still *UP* with probability $1 - \beta$, that is

$$\mathrm{res_{pr}}(\mathbf{N}, \beta) = \max \left\{ I : \sum_{i=1}^{I} P(\mathbf{N}, i) \leq \beta \right\}.$$

In ref. [3], network components are the nodes, and the authors assume that the node failures appear in *random order*, that is all $n!$ node orderings are equally probable. Obviously similar definitions remain valid in the case when the components subject to failure are the network links.

1.3 D-Spectra

1.3.1 Introduction

In this section, we introduce a topological invariant of the network which we call multidimensional destruction spectrum (D-spectrum). The D-spectrum will be our main tool for studying network with more than two states.

Let us assume that the network can be in several states which we formally mark by integer J. $J = K$ corresponds to the "best" state with highest performance level of the network. $J = K - 1, K - 2, \ldots$, denote network states in the process of its gradual deterioration (disintegration). State $J = 0$ is assigned to network total failure (collapse).

For example, consider a network in which all nodes are terminals. The components of the network subject to failure are the links. Assume that links fail one after another, in random order. Network state is defined in accordance with the number of isolated clusters in it. Initially, there is a single cluster, and this is the best state of the network. Suppose that the maximal permissible number of clusters which allows the network to operate is 3. When network falls apart into 4 or more clusters it is considered as its total failure (collapse). So, we define network state $J = 3$ when there is exactly one cluster, $J = 2$—for two clusters, $J = 1$—for three clusters and finally $J = 0$ (*DOWN*) for four or more clusters.

When network components which fail are the links, a single link failure either does not change the number of clusters in the network or causes its increase by one. Similar is the situation with the network whose state is defined as the maximal s–t flow when links have capacity equal one.

The situation is different when the network components subject to failure are *nodes*. Consider, for example, the network shown on Fig. 1.3b. It has five nodes subject to failure. Let us remind that node failure means elimination of all links incident to this node while the node remains intact. Let the state of the network be defined as [5—*the number of isolated components*]. If all 5 nodes are *up*, network state is $J = 4$ because all nodes constitute a single component. Suppose the nodes fail in the following order: $\{1, 2, s\}$. After the first failure we have two isolated components and therefore $J = 3$. (An isolated node is also considered a component). After node 2 fails, network state becomes $J = 2$. Finally, after node s fails, all nodes become isolated, and the network state drops to $J = 0$ (*DOWN*). So, in this example, a single component failure causes network state change by *two* units.

Formally, it is convenient to distinguish between two types of networks: *single-step* and *multi-step* .

Definition 1.3.1 (*Single-step network*)
If a single component failure either leaves the network state unchanged or causes its decrease by one, the network is called *single-step*. Otherwise, we call it *multi-step*.

1.3.2 D-Spectrum of a Single-Step Network

Denote by $e_1, e_2, \ldots, e_n$ the network components which are subject to failure. Suppose that the network can be in $K + 1$ states numbered $J = K$, $J = K - 1, \ldots$, $J = 1, J = 0$. Consider a random permutation π of network component numbers:

$$\pi = (e_{i_1}, e_{i_2}, e_{i_3}, \ldots, e_{i_n})$$

Suppose that all these components are *up* and we move along the permutation, from left to right, and turn each component from *up* to *down*. Suppose, the network state is controlled after each step. In a single-step network, we will observe exactly K occasions when the network state has changed by one, from K to $K - 1$, from $K - 1$ to $K - 2$, etc, until the entrance from state $J = 1$ into the state $J = 0$ (*DOWN*).

Definition 1.3.2
(*The anchors*) The ordinal number in the permutation π of the component whose turning *down* causes network state to change from $J = K - I$ to $J = K - I - 1$, $I = 0, 1, \ldots, K - 1$, is called the $(I + 1)$th *anchor* and is denoted r_{I+1}. Each permutation has therefore K anchors.

Example 1.3.1 (*Link failures in four terminal network*, Fig. 1.4)
The figure shows the gradual disintegration of the four-terminal network when the links fail in the following random order $\pi = (1, 3, 4, 2)$. The network has four states: $J = 3$, when there is one cluster, $J = 2$ for two clusters, $J = 1$ for three clusters and $J = 0$ (*DOWN*) for four clusters. Correspondingly, the transition $3 \to 2$ occurs when link 3 fails, $2 \to 1$—when link 4 fails and $1 \to 0$ when link 2 fails. Therefore, the first anchor equals 2, the second—3, and the third anchor is 4. Note that anchor value is *not* the link number but the *ordinal* number of the *position* of the corresponding link in the random permutation.

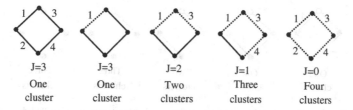

Fig. 1.4 Disintegration of four terminal network. Links fail in the following order: $\pi = (1, 3, 4, 2)$

Assume that all permutations are equally probable. Each permutation of n components has probability $1/n!$.

Definition 1.3.3 (*Multidimensional D-spectrum*) The K-dimensional discrete density

$$f(\alpha_1, \alpha_2, \ldots, \alpha_K) = P\ (r_i(\pi) = \alpha_i,\ i = 1, 2, \ldots, K)$$
$$= \frac{\text{number of permutations with } r_i(\pi) = \alpha_i,\ i = 1, 2, \ldots, K}{n!} \qquad (1.3.1)$$

for $1 \le \alpha_1 < \alpha_2 < \cdots < \alpha_K \le n$ is called network *multidimensional D-spectrum*.

A few comments. Letter "D" for the spectrum signifies the process of network *destruction* since we eliminated (turned from *up* to *down*) its components moving along the permutation from left to right. Later we will give examples of a dual procedure when we move from right to left along π and turn components from *down* to *up*. It will produce a C-spectrum, "C" stands for *construction*.

Obviously, for single-step networks,

$$\sum_{1 \le \alpha_1 < \alpha_2 < \cdots < \alpha_K \le n} f(\alpha_1, \alpha_2, \ldots, \alpha_K) = 1.$$

It is important to stress that the D-spectrum is a *combinatorial parameter* of the network. It depends only on the network structure and the definition of its states. It does not depend on probabilistic characterization of the real random mechanism which governs network component failures. In particular, if the network components have i.i.d. lifetimes with continuous CDF $G(t)$, the D-spectrum remains the same for any $G(t)$.

Let us consider several examples. First, return to the network on Fig. 1.4. In the permutation $\pi = (1, 3, 4, 2)$, $r_1 = 2, r_2 = 3, r_3 = 4$. Due to the symmetry of the network, the same positions of the anchors will be in every one of $4! = 24$ permutations. Therefore for this example, for each π, always $\alpha_1 = 2, \alpha_2 = 3, \alpha_3 = 4$ and

$$f(r_1 = 2, r_2 = 3, r_3 = 4) = 1.$$

More interesting example is presented on Fig. 1.5.

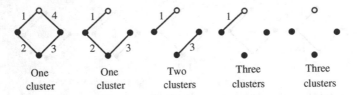

One   One   Two   Three   Three
cluster   cluster   clusters   clusters   clusters

Fig. 1.5 Network with four links subject to failure and three terminals (*bold*). Edges fail in the order $\pi = (4, 2, 3, 1)$

Example 1.3.2 (Network with four links and three terminals)
Here the network has three states: $J = 2$ for one cluster, $J = 1$ for two clusters and $J = 0$ (*DOWN*) for three clusters. Links are subject to failure. Correspondingly, there are two anchors, r_1, r_2. The first corresponds to the transition $2 \rightarrow 1$, the second—to the transition $1 \rightarrow 0$. For the permutation $\pi = (4, 2, 3, 1)$, $r_1(\pi) = 2, r_2(\pi) = 3$. Analyzing all $4! = 24$ permutations we can find out that in 12 of them we have $r_1(\pi) = 2, r_2(\pi) = 3$, in 4 of them $r_1(\pi) = 3, r_2(\pi) = 4$, and in remaining 8—$r_1(\pi) = 2, r_2(\pi) = 4$. Therefore,

$$f(r_1 = 2, r_2 = 3) = 12/24, \quad f(r_1 = 3, r_2 = 4) = 4/24,$$

$$f(r_1 = 2, r_2 = 4) = 8/24.$$

Our main interest will be the probabilistic description of each particular anchor. More formally, we will be interested in the *marginal* distribution of the position of each of the K anchors.

Definition 1.3.4 (*The sth marginal D-spectrum*)
The distribution

$$\mathbf{f}^{(s)} = \left(f_1^{(s)}, f_2^{(s)}, \ldots, f_n^{(s)} \right)$$

of the position of the sth anchor is called the sth marginal D-spectrum. Here

$$f_i^{(s)} = P \text{ (the sth anchor position is } i) \,. \tag{1.3.2}$$

Obviously,

$$f_i^{(s)} = P(r_s(\pi) = i)$$

$$= \sum_{\{1 \le \alpha_1 < \cdots < \alpha_s = i < \cdots < \alpha_K \le n\}} f(\alpha_1, \alpha_2, \ldots, \alpha_s = i, \ldots, \alpha_K) \,. \tag{1.3.3}$$

Example 1.3.2-continued
It is easy to find out that

$$\mathbf{f}^{(1)} = (0, 20/24, 4/24, 0), \, \mathbf{f}^{(2)} = (0, 0, 12/24, 12/24).$$

It is more convenient to operate with so-called *cumulative* (marginal) spectrum.

Definition 1.3.5 (*The sth cumulative D-spectrum*)
The cumulative distribution function $F^{(s)}(x)$ of the position of the sth anchor in random permutation π is called the sth *cumulative D-spectrum*:

$$F^{(s)}(x) = \sum_{i=1}^{x} f_i^{(s)}, \quad x = 1, 2, \ldots, n. \tag{1.3.4}$$

Let us clarify the probabilistic meaning of $F^{(s)}(x)$. Divide all network states $\{J\}$ into two sets, U and D. To set U belong all states $J : J > K - s$, and to the complementary set D all the remaining states $J : J \leq K - s$. Denote by $Y_{(s)}$ the random number of components needed to be turned *down* in the course of the destruction process to cause the transition from U to D.
Obviously,

$$f_i^{(s)} = P\left(Y_{(s)} = i\right), \quad i = 1, \ldots, n.$$

Then

$$F^{(s)}(x) = P\left(Y_{(s)} \leq x\right).$$

In words: $F^{(s)}(x)$ is the CDF of the number of components to be destroyed to cause the transition from U to D.

Example 1.3.2-continued
It is easy to find out that

$$F^{(1)}(1) = 0, \; F^{(1)}(2) = 5/6, \; F^{(1)}(3) = 1 = F^{(1)}(4).$$

Similarly,

$$F^{(2)}(1) = 0, \; F^{(2)}(2) = 0, \; F^{(2)}(3) = 1/2, \; F^{(2)}(4) = 1.$$

Remark 1 Suppose that the network has four mutually exclusive states:

$$UP, DOWN1, DOWN2, DOWN3.$$

On each particular permutation π the first anchor signifies the transition from $UP \rightarrow DOWN1$, the second anchor—the transition $DOWN1 \rightarrow DOWN2$ and the third anchor—the transition $DOWN2 \rightarrow DOWN3$. Suppose x components have failed in random order. Then $F^{(1)}(x)$ is the probability that the network is in $DOWN1$ *or* $DOWN2$ *or* $DOWN3$. $F^{(2)}(x)$ is the probability that the network is in state $DOWN2$ *or* $DOWN3$. And $F^{(3)}(x)$ is the probability that the network is in state $DOWN3$. This reasoning is very similar to the claim that if the rth event in the renewal process took place in the interval $[0, t_0]$, then there were r *or more* events on the interval $[0, t_0]$.

Now let us describe the principal combinatorial property of the cumulative marginal D-spectra. For this purpose, let us consider a *binary* network, i.e. a network with two states $J = 1$ (*UP*) and $J = 0$ (*DOWN*). If a network has more than two states, we divide them into two groups $J \geq L$ and $J < L$. For example, if the network has four states $J = 3, 2, 1, 0$ we take $L = 2$ and declare states $J = 3, 2$ as the *UP* state and the remaining states $J = 1, 0$ as *DOWN* state. Here the *second* marginal D-spectrum determines the probability $f_k^{(2)}$ that the transition from *UP* to *DOWN* takes place after turning *down* the kth component in the permutation. To simplify the notation we will omit the upper index s at the relevant marginal D-spectra.

Assume now that the network is in the *DOWN* state. Consider the set S_{DOWN} of all vectors $\mathbf{v} = (x_1, x_2, \ldots, x_n)$ with binary 0/1 components such that $\mathbf{v} \in S_{DOWN} \Leftrightarrow \varphi(\mathbf{v}) = 0$. $\mathbf{v}$ is called failure vector (also a cut vector, see Definition 1.1.2). The set S_{DOWN} can be divided into several subsets according to the number of zeroes in the failure vectors. Denote by $C(x)$ the number of failure vectors with exactly x zeroes, $x = 1, 2, \ldots, n$.

The following theorem establishes an important combinatorial property of the cumulative D-spectrum .

Theorem 1.3.1

$$C(x) = F(x) \cdot \frac{n!}{x!(n-x)!}, \quad x = 1, 2, \ldots, n. \tag{1.3.5}$$

Before proving (1.3.5), let us illustrate Theorem 1.3.1 by an example.

Example 1.3.3 (Failure sets of bridge network, Fig. 1.3a)
The bridge failure is defined as disconnection of s and t. The bridge never fails when only one link is down. Therefore, $f_1 = 0$. Analyzing $5! = 120$ permutations of 5 links, it is easy to find out that the bridge fails when exactly two links fail in 24 of them, and thus $f_2 = 1/5$. After four links are down, the bridge is always down. Thus, $f_5 = 0$. There are exactly 24 permutations with the anchor on the fourth position. Thus, $f_4 = 1/5$. Therefore, $f_3 = 3/5$, and

$$F(1) = 0, \; F(2) = 1/5, \; F(3) = 4/5, \; F(4) = 1, \; F(5) = 1.$$

By the Theorem 1.3.1, $C(1) = 0, C(2) = 2, C(3) = 8, C(4) = 5, C(5) = 1$. Indeed, there are no failure vectors with only one component *down*, there are exactly two failure vectors with two components *down*, namely $(0,0,1,1,1)$ and $(1,1,1,0,0)$, and 8 failure vectors with exactly three zeroes, etc.

Proof ([10], p. 114)
Consider a random permutation $\pi = (i_1, \ldots, i_n)$ Declare the first x elements of it as being *down* and the rest as being *up*. If this permutation determines the network *DOWN* state, call it $(x; D)$-permutation. Denote by $N(x)$ the total number of (x, D)-permutations. Obviously, the probability to have an (x, D)-permutation equals $N(x)/n!$. But on the other hand, this probability equals $F(x) = f_1 + \cdots + f_x$ by our definition of the destruction process. Therefore, $F(x) = N(x)/n!$. When we

define a *DOWN* state with exactly x components being *down*, their location in π is not relevant. All permutations obtained by permuting x *down* components between themselves and the remaining $(n - x)$ between themselves correspond, in fact, to the same failure vector with x components being *down*. Therefore, $C(x) = N(x)/(x!(n - x)!) = F(x) \cdot n!/(x!(n - x)!)$, which completes the proof. $\qquad\square$

Suppose we chose *randomly* a set of x components in the network and declare them *down*. All sets of x components chosen from n components are equally probable. There are, as we have denoted, $C(x)$ sets with x zeroes that are failure sets. Then it follows from the theorem above that the probability that such set will cause network failure is $C(x)$ divided by the total number of ways to select x components out of n, i.e. it equals $F(x)$. This property of the cumulative D-spectrum was used in [14] to estimate the damage associated with network destruction which is subject to an attack on its links.

Remark 2 Let us now recall the Definition 1.2.3 of probabilistic resilience. Obviously, $1 - F(x)$ is the probability that the network is *UP* after turning x of its components (nodes or vertices) to *down*. Let $x_{\max}$ be the largest x such that $1 - F(x) > 1 - \beta$. Then this $x_{\max}$ is the probabilistic β-resilience of the network according to the Definition 1.2.3.

Remark 3 The networks with many states and their D-spectra have been introduced in [14] and [11], see also [12]. The connection between the D-spectra and system failure states (1.3.5) is a known fact. For example, F. Samaniego [19] mentions the connection between $S(x) = 1 - F(x)$ and the number of systems *path sets* with exactly x components in *up* state, which is, in fact, a dual equivalent of (1.3.5).

Remark 4 In applications we operate with the D-spectra in the form of the marginal *discrete density* $\{f_i\}$ or in the form of the *marginal cumulative* D-spectrum $\{F(x)\}$. For sake of brevity, we often omit the words "marginal" and/or "cumulative", and rely on the notation $f.$ and $F(\cdot)$.

1.3.3 Formula for Network DOWN Probability

Suppose that the network has two states, *UP* and *DOWN* and we know its cumulative marginal D-spectrum (1.3.4) $F(x)$, $x = 1, 2, \ldots, n$. Suppose that the network has independent binary components, and the probability that the ith component is *up* equals p, the same for all i. Denote by $Q_N(p)$ the probability that the network is *DOWN* and let $q = 1 - p$. The connection between $F(x)$ and failure set with $C(x)$ established by Theorem 1.3.1 leads us to the formula for $Q_N(p)$.

Corollary 1.3.1 (*Formula for $Q_N(p)$.*)

$$Q_N(p) = \sum_{x=1}^{n} C(x)q^x p^{(n-x)} = \sum_{x=1}^{n} F(x)q^x p^{(n-x)} \frac{n!}{x!(n - x)!}. \qquad (1.3.6)$$

Proof Each failure set with x components *down* and $(n - x)$ components *up* has probability $q^x p^{(n-x)}$. All failure sets with x component *down* have probability $C(x)q^x p^{(n-x)}$. If the network is *DOWN*, it must be in one of its down states. This explains the sum in (1.3.6). □

The Corollary allows another interpretation in terms of network *lifetime*. Suppose that all network components have i.i.d. random lifetimes $\tau_1, \tau_2, \ldots, \tau_n$, $P(\tau_i \leq t) = G(t)$ for all $i = 1, \ldots, n$. Assume that $G(t)$ is a continuous function. Suppose all components start functioning at time $t = 0$. Consider a time instant t_0. Let $P(\tau_i \leq t_0) = q$. q is the probability that component i is *down* at the instant t_0. (Formally, we should write at the instant $t_0 + 0$). Therefore, the expression

$$\sum_{x=1}^{n} C(x)(G(t_0))^x (1 - G(t_0))^{(n-x)} = P(\text{network is } DOWN \text{ at } t_0).$$

But the last expression gives the probability that network lifetime $\tau_N \leq t_0$. Denoting network lifetime CDF by $G_N(t_0)$, we arrive at the expression

$$G_N(t_0) = \sum_{x=1}^{n} C(x)(G(t_0))^x (1 - G(t_0))^{(n-x)}. \tag{1.3.7}$$

The expression for the mean network lifetime is the following:

$$E[\tau_N] = \int_0^\infty (1 - G_N(t)) \, dt. \tag{1.3.8}$$

What can be done if not all components are identical? Formally, (1.3.6) is not valid if not all $p_i \equiv p$. Nevertheless, this formula can be useful if it is known that all p_i lie in a relatively narrow interval $[p_{min}, p_{max}]$. It is a well-known fact established in reliability theory, see e.g. [1] , that system reliability is a monotone increasing function of component reliability. Therefore, for $p_i \in [p_{min}, p_{max}] = \Delta$,

$$Q_N(p_1, \ldots, p_n) \in [Q_N(p_{max}), Q_N(p_{min})]. \tag{1.3.9}$$

This formula might be useful if our information about component reliability is not accurate but allows to establish a relatively narrow interval Δ for p_i values.

Remark 1

The formula (1.3.7) gives the CDF of network lifetime for the case of i.i.d. component lifetimes $\tau_i \sim G(t)$. This formula is equivalent to the well-known Samaniego's representation [18, 19] of coherent system lifetime via the so-called system *signature* and the CDF's of order statistics of the random sample $\tau_1, \tau_2, \ldots, \tau_n$. Numerically the signature coincides with the D-spectrum of a binary network.

In Samaniego's definition of signature, all components of the system have i.i.d. continuous lifetime distributions. Equation 1.3.9 allows to obtain simple bounds on system lifetime distribution for the case when the i.i.d. condition is replaced by weaker condition of independent continuous component lifetimes which may not be necessarily identical.

Suppose that component i has a continuous lifetime $\tau_i \sim G_i(t)$. Define the *upper* and the *lower* CDF's $G_U(t)$ and $G_L(t)$, respectively, according to the formulas:

$$G_U(t) = \max [G_1(t), G_2(t), \dots, G_n(t)], t \in (0, \infty),$$

and

$$G_L(t) = \min [G_1(t), G_2(t), \dots, G_n(t)], t \in (0, \infty).$$

Then obviously, for each $t^\star$, component failure probability $q(t^\star) \in [G_L(t^\star), G_U(t^\star)]$. Using (1.3.7) we immediately obtain bounds on network lifetime CDF in the form

$$G_{\mathbf{N}}(t^\star) \in [G_{\mathbf{N}}^{(L)}(t^\star), G_{\mathbf{N}}^{(U)}(t^\star)],$$

where

$$G_{\mathbf{N}}^{(L)}(t^\star) = \sum_{x=1}^{n} C(x)(G_L(t^\star))^x \left(1 - G_L(t^{(\star)})\right)^{(n-x)}$$

and

$$G_{\mathbf{N}}^{(U)}(t^\star) = \sum_{x=1}^{n} C(x)(G_U(t^\star))^x \left(1 - G_U(t^{(\star)})\right)^{(n-x)}.$$

More detailed discussion on F.Samaniego's approach will be continued in Sect. 1.3.5.

Let us conclude this section by a formula which gives an approximation to $Q_{\mathbf{N}}(p)$ for the case of a highly reliable network. Formally, let us assume that $q = (1 - p) = \delta \to 0$. Then the main term in the formula (1.3.6) will be the first term with minimal nonzero $C(x) = C(x_{\min})$. Its contribution will be $C(x_{\min})\delta^{x_{\min}}$ and

$$Q_{\mathbf{N}}(p = 1 - \delta) = C(x_{\min})\delta^{x_{\min}}(1 + o(1)) \tag{1.3.10}$$

as $\delta \to 0$.

Note that $x_{\min}$ is the size of the minimal-size min cut set of the network. For example, in the bridge-type s-t network there are two cut sets of size two, i.e. $x_{\min} = 2$, and the main term will be $2\delta^2$. Later we will demonstrate that (1.3.10) gives quite accurate failure probability estimates for small δ. We call the main term in (1.3.10) the Burtin–Pittel approximation, see [8] and reference there. This approximation was first suggested in a slightly different form by Burtin and Pittel [4] in 1972.

1.3.4 Networks with Many States

Now let us return to the situation where the network has $K + 1$ states, denoted as $J = K$, $J = K - 1, \ldots,$ $J = 1$ and $J = 0$. As a typical example we can have in mind a network with $h = K + 1$ terminals. State $J = K$ means that all terminals are connected to each other, $J = K - 1$ denotes the situation with two isolated clusters, and so on. Finally, $J = 0$ means complete network collapse, i.e. its disintegration into $K + 1$ isolated clusters. Let p be the probability that the network component is *up*, the same for all components. Assume also that components fail independently.

Our goal is to derive the formulas for $P_N(J; p)$, the probability that the network is in state $J = K$, $K - 1, \ldots, 0$. This will be done by means of the marginal D-spectra, using the dichotomy of network states into two groups: the states with $J \geq L$ and the complementary group with $J < L$, where $L = K$, $K - 1, \ldots, 1$.

Let us declare the state $J = K$ as network *UP* state and all other states with $J < K$ as network *DOWN* state. Then, using (1.3.6) and our first marginal D-spectrum $F^{(1)}(x)$ we can write that the probability that **N** is *DOWN*, (i.e. is in one of the states $J = K - 1, \ldots, J = 0$) is equal to

$$\sum_{J=1}^{K-1} P_N(J; p) = \sum_{x=1}^{n} F^{(1)}(x) q^x p^{(n-x)} \frac{n!}{x!(n-x)!}. \qquad (1.3.11)$$

Next let us declare the states $J = K$ and $J = K - 1$ as our new *UP* state and all the remaining states—as the new *DOWN* state. Then, using our second marginal D-spectrum we obtain that

$$\sum_{J=1}^{K-2} P_N(J; p) = \sum_{x=1}^{n} F^{(2)}(x) q^x p^{(n-x)} \frac{n!}{x!(n-x)!}. \qquad (1.3.12)$$

Comparing (1.3.11) and (1.3.12) we obtain that

$$P_N(K - 1; p) = \sum_{x=1}^{n} \left[F^{(1)}(x) - F^{(2)}(x) \right] q^x p^{(n-x)} \frac{n!}{x!(n-x)!}. \qquad (1.3.13)$$

Moving further the "barrier" between the *UP* and *DOWN* states, we arrive at the desired result formulated as

Theorem 1.3.2 *Put $F^{(K+1)}(x) = 0$. Then for $J = K - 1, \ldots, 0$,*

$$P_N(J; p) = \sum_{x=1}^{n} \left[F^{(K-J)}(x) - F^{(K-J+1)}(x) \right] q^x p^{(n-x)} \frac{n!}{x!(n-x)!}. \qquad (1.3.14)$$

1.3.5 D-Spectrum and Signature

The readers familiar with Reliability theory and with so-called *signatures* of coherent systems, see e.g. [17–19], may realize that numerically the signature $\mathbf{s} = (s_1, s_2, \ldots, s_n)$ coincides with the marginal D-spectrum of binary network $\mathbf{f} = (f_1, f_2, \ldots, f_n)$ introduced in Sect. 1.3.2.

F. Samaniego [19], p. 21 gives the following definition:

> Assume that the lifetimes of coherent systems n components are independent and identically distributed according to the (continuous) distribution G. The signature denoted by $\mathbf{s}$ is an n-dimensional probability vector whose ith element s_i is equal to the probability that ith component failure causes the system to fail. In brief, $s_i = P(T = X_{i:n})$, where T is the failure time of the system and $X_{i:n}$ is the ith order statistic of the n component failure times, that is, the time of ith component failure.

The principal result of F. Samaniego [18, 19] is the representation of the system lifetime distribution $G_s(t)$ in the following form:

$$G_s(t) = P(T \le t) = \sum_{i=1}^{n} s_i \, P(X_{i:n} \le t). \qquad (1.3.15)$$

If we substitute into this formula the well-known expression of the CDF of ith order statistic, see [5], and rearrange the terms, we arrive at the expression (1.3.6) for network failure probability.

Suppose that the network $\mathbf{N}$ has three states, $J = 2$, $J = 1$ and $J = 0$ (*DOWN*). Let the network starts its life at time $t = 0$ in state $J = 2$. Denote by $\tau_{N,1}$ the entrance time into state $J = 1$, and by $\tau_{N,2}$ the network lifetime, i.e. the time of entrance into $J = 0$. In our notation, the two-dimensional D-spectral density $f(i, j)$ is the probability that the transitions $J = 2 \rightarrow J = 1$ and $J = 1 \rightarrow J = 0$ coincide with the ith and jth order statistics, respectively, of the random sample $X_1, X_2, \ldots, X_n$, $1 \le i < j \le n$. Using the Law of Total Probability, we arrive at the expression:

$$P\left(\tau_{N,1} \le T_1 \,\&\, \tau_{N,2} \le T_2\right) = \sum_{1 \le i < j \le n} f(i, j) G_{(i,j)}(T_1, T_2), \qquad (1.3.16)$$

where $G_{(i,j)}(T_1, T_2)$ is the joint CDF of (i, j)th order statistics, see [5], p. 11. This formula is a two-dimensional analogue of Samaniego's formula (1.3.15).

The recent work [17] calls f_i in (1.3.2) the *discrete signature*. The work [6] introduced the D-spectrum under the name *Internal Distribution* (ID) and presented Monte Carlo simulated ID's for several complete graphs and the dodecahedron network. Later on, instead of ID we used the term D-spectrum, see [9, 10, 12].

Let us return to the properties of the cumulative D-spectrum, Sect. 1.3.2. We have established that the cumulative D-spectrum is directly connected to the number of failure sets of the given structure (network), see formula (1.3.5). Speaking formally, the D-spectrum is a functional of the system structure function. If the structure function $\varphi(\mathbf{x}) \rightarrow \{0, 1\}$ is known, we can obtain the D-spectrum. If so, we can say that the signature defined above by Samaniego [18, 19] is in fact a system structural characteristic which can be obtained without any assumptions regarding the component lifetime distribution, and in particular, without the i.i.d. assumption.

The i.i.d. assumptions were critical to establish the principal formula (1.3.15) which gives us the the *system* lifetime CDF $G_s(t)$ representation via the system signature. This formula, after simple algebra, can be rewritten in the following more compact form:

$$G_s(t) = \sum_{x=1}^{n} S(x) \cdot (G(t))^x (1 - G(t))^{n-x} n!/(x!(n-x)!) \qquad (1.3.17)$$

where $G(t)$ is the component lifetime CDF, and $S(x) = s_1 + s_2 + \cdots + s_x$ is the Samaniego's cumulative signature (coinciding with the cumulative D-spectrum $F(x) = f_1 + \cdots + f_x$).

Let us try to answer the following question:

Which properties of $G_s(t)$ are preserved, partially or completely, after some of the i.i.d. and G-continuity assumptions will be relaxed?

Let us reexamine (1.3.17) and put $t = t_0$. Then we have the system *DOWN* probability $G_s(t_0)$ at the instant t_0 in the following form:

$$G_s(t_0) = \sum_{x=1}^{n} S(x) q_0^x (1 - q_0)^{(n-x)} n!/(x!(n-x)!), \qquad (1.3.18)$$

where $q_0 = G(t_0)$. We see, therefore, that $G_s(t_0)$ depends only on the value $G(t)$ at the point $t = t_0$. The information about the values of $G(t)$ for other t values is *not* relevant. Note that *independence* of components is important because without it we can not determine the probabilistic weight of a failure set with x components being *down* and $(n - x)$ being *up* as $q_0^x (1 - q_0)^{(n-x)}$.

Now assume that all we know about $G(t_0) = q_0$ is that it lies within an interval $\delta = [q_{min}, q_{max}]$. Then by the monotone property of the reliability function, see [1], Chap. 1, we will obtain that

$$G_s(t_0) \in \Delta = [G_s(q_{min}), G_s(q_{max})]. \qquad (1.3.19)$$

Now imagine that a single component j is different from others, and its CDF is $G_j(t_0) = q_j$, $q_{min} < q_j < q_0$. Then, by the same monotone property of reliability function, the failure probability of the system will become smaller but remains in the same interval Δ. Now it is obvious that component j lifetime CDF's $G_j(t)$, $j = 1, \ldots, n$, can be *different*, and (1.3.19) holds true if all $G_j(t_0) \in \delta$.

We arrive therefore at the conclusion that (1.3.19) holds if system components are independent but *not necessary identically distributed*.

The reasoning above can be repeated by moving the point t_0 along the t-axis. Suppose that the component lifetime CDF's $G_j(t)$, $j = 1, 2, \ldots, n$, lie within a "belt" created by their supremum and infimum. So, suppose, that there are such two continuous CDF's $G_L(t)$ and $G_U(t)$ that for all $t \geq 0$

$$G_L(t) \le G_j(t) \le G_U(t), \quad j = 1, 2, \ldots, n.$$

Then, obviously, (1.3.19) can be modified as follows.

$$G_s(t) \in [G_L(t), G_U(t)], \tag{1.3.20}$$

where

$$G_L(t) = \sum_{i=1}^{n} s_i P\left(Y_{(i:n)}^L \le t\right), \tag{1.3.21}$$

and

$$G_U(t) = \sum_{i=1}^{n} s_i P\left(Y_{(i:n)}^U \le t\right). \tag{1.3.22}$$

Here $Y_{(i:n)}^L$ and $Y_{(i:n)}^U$ are the corresponding order statistics from the populations with CDFs $G_L(t)$ and $G_U(t)$, respectively. Summing up, we have relaxed the assumption of *identical* distribution of component lives but for the price that system lifetime CDF will not be known exactly and will lie within bounds.

The assumption that all component lifetimes $G_j(t)$, $j = 1, \ldots, n$, must be continuous can be relaxed too. We will not go into this rather technical issue.

Remark The number of isolated clusters in the network is not the only way to define the number of states in the multistate case. Another approach of practical importance to defining the state of the network is the follow-up of the *size of the maximal isolated component* in the network determined as the number of nodes in this component. Consider, for example, the bridge network, Fig. 1.3a. It has four nodes and five links. Suppose the links fail in our destruction process in the following order: 1, 5, 2, 3, 4. The size of the maximal connected component is $x_{max} = 4$ (after links 1 and 5 fail), then it becomes 3, 2 and 1, respectively, after the failure of links 2, 3 and 4.

An example which uses this type of network state definition will be considered in Sect. 2.1. We will define three states of 32-node network which will be subject to a random "attack" on its nodes: $x_{max} > 10$, $3 < x_{max} \le 10$, and $x_{max} \le 3$.

1.3.6 Renewal Process of Component Failures

There might be several approaches to modeling the process of network component failures. The simplest one is to assume that the components fail according to independent random "lotteries" in which component i is declared to be *up* or *down* with probability $p_i \equiv p$ and $1 - p_i \equiv q$, respectively. Samaniego's model [18] puts the component failures into a frame of a process developing in time by assuming that the rth failure time equals to $X_{r:n}$, the rth order statistic from the sample of i.i.d.

component lifetimes $X_1, X_2, \ldots, X_n$. The i.i.d. and G-continuity assumption guarantees that all orders of component failure appearance in time are equally probable. A natural probabilistic model for the process of component failure appearance might be the following: component failures appear according to a renewal process $\xi(t)$ defined as a sequence of i.i.d. positive random variables $Z_1, Z_2, \ldots, Z_k, \ldots$, see [1], Chap. 6. The events in $\xi(t)$ appear at the instants $\vartheta_k = \sum_{m=1}^{k} Z_m, k = 1, 2, \ldots$. Assume also that all orders of component failures appearance are equally probable. Then a natural variation of formula (1.3.15) may be derived as follows.

Suppose the network has two states, *UP* and *DOWN*, and we know the corresponding cumulative D-spectrum $F(x)$. Denote by $N(t)$ the number of renewals in the interval $[0, t]$. Formulas for $P(N(t) = k) = \rho_k$ are well known in Renewal Theory and easily obtainable via the convolutions of the CDFs of r.v.'s Z_i. On the other hand, we know the probability $F(x)$ that the network is *DOWN* if x of its components are turned *down*. Let T be the network lifetime. Therefore, by the formula of Total Probability,

$$P(T \leq t) = \sum_{x=1}^{\infty} \rho_x \cdot F(x),$$

where we set $F(x) \equiv 1$ for $x > n$.

1.4 Series, Parallel and Recurrent Systems

1.4.1 Spectra of Parallel, Series and Recurrent Networks

Suppose we have two s–t networks N_1 and N_2. We know the cumulative D-spectra of these networks, $F_1(x), x = 1, 2, \ldots, n$ and $F_2(y), y = 1, 2, \ldots, m$. We create a new network N which is a parallel connections of N_1 and N_2. This situation can take place, for example in the design of communication networks. Our goal is to find out the D-spectrum of N. It will be assumed that both networks consist of independent and identical components, i.e. $q_1 = q_2 = q$.

The material below is based on the theory and ideas presented in detailed and general form in the recent paper [21].

By the definition of parallel system, N is *DOWN* if and only if both N_1 and N_2 are *DOWN*.

$$P(\mathbf{N} \text{ is } DOWN) = \left(\sum_{x=1}^{n} C_1(x) q^x p^{n-x} \right) \cdot \left(\sum_{y=1}^{m} C_2(y) q^y p^{n-y} \right), \qquad (1.4.1)$$

where $C_1(x), C_2(y)$ are the number of failure sets of size x and y in the first and the second network, respectively. After simple algebra, this formula takes the form

Fig. 1.6 Series connection
of two small networks (**a**)
and a recurrent system (**b**)

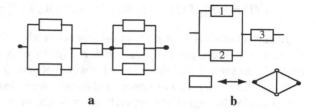

a   b

$$P(\mathbf{N} \text{ is } DOWN) = \sum_{z=1}^{n+m} C(z) q^z p^{n+m-z}, \qquad (1.4.2)$$

where

$$C(z) = \sum_{j=1}^{z} C_1(j) C_2(z - j), \quad z = 1, 2, \ldots, n + m.$$

From here it follows that the D-spectrum of the parallel network **N** is

$$F_{\mathbf{N}}(z) = C(z) \cdot z!(n + m - z)!/(n + m)!, \quad z = 1, \ldots, n + m. \qquad (1.4.3)$$

Very similar is the situation with two networks in series connection. We will consider this situation on an example, see Fig. 1.6a.

Example 1.4.1 (Two networks in series)
Both networks have three components, see Fig. 1.6a. The first network has $C_1(1) = 1$, $C_1(2) = 3$, and $C_1(3) = 1$, and the second—only one failure set of size 3, i.e. $C_2(3) = 1$. Let Q_1 and Q_2 be the *DOWN* probabilities of the first and second network, respectively. Then the *DOWN* probability for the series connection of these networks is

$$Q_{\mathbf{N}} = 1 - (1 - Q_1)(1 - Q_2) = 1 - \left(1 - qp^2 - 3q^2p - q^3\right)(1 - q^3).$$

Now it remains to bring the expression for $Q_{\mathbf{N}}$ to the (qp)-polynomial form. This will be done by replacing 1 by $(q + p)^6$ and by replacing 1 by $(q + p)^3$ in the terms $(1 - Q_1)$ and $(1 - Q_2)$:

$$Q_{\mathbf{N}} = (q + p)^6 - \left((q + p)^3 - Q_1\right)\left((q + p)^3 - Q_2\right).$$

Now apply the operator $Expand[Q_{\mathbf{N}}]$ of *Mathematica* [22] to the expression of $Q_{\mathbf{N}}$. The output will be the following:

$$Q_{\mathbf{N}} = qp^5 + 6q^2p^4 + 14q^3p^3 + 15q^4p^2 + 6q^5p + q^6.$$

We obtain, therefore, that the network has $C(1) = 1$, $C(2) = 6$, $C(3) = 14$, $C(4) = 15$, $C(5) = 6$, $C(6) = 1$. From (1.4.3) it follows that the cumulative D-spectrum of the series network is

$$F(1) = 5/30, \; F(2) = 12/30, \; F(3) = 21/30, \; F(4) = 1 = F(5) = F(6).$$

Recurrent network is an arbitrary s–t network **N** whose links are *identical* arbitrary s–t networks. More precisely, each link $e = (a, b)$ of **N** is replaced by an $s_1 - t_1$ network, the same for all links, in such a way that node $a : \; = s_1$ and $b : \; = t_1$.

N is called the *organizing* network and its links are called *modules*, see [1], Chap. 1. For example, the organizing network might be the network shown on Fig. 1.1 and each of its eleven modules may be an s–t network, in particular case, a bridge. We assume that all modules are identical and have independent and identical components.

Following Sect. 4 of [21], we will obtain the D-spectrum of a recurrent network in the following way.

Assume that we know the D-spectra of the organizing network and of the module network.

Let the formula for the *DOWN* probability of the organizing structure **N** be as follows:

$$P(\mathbf{N} \text{ is } DOWN) = \sum_{x=1}^{n} C^{\star}(x)(q^{\star})^{x}(p^{\star})^{n-x}, \tag{1.4.4}$$

where $C^{\star}(x)$ is the number of failure sets of size x which is expressed via the D-spectrum $F^{\star}(x)$ of the organizing network as

$$C^{\star}(x) = F^{\star}(x) \cdot n!/(x!(n-x)!). \tag{1.4.5}$$

In (1.4.4), $q^{\star}$ is the *down* probability of the module, which has an expression:

$$q^{\star} = \sum_{y=1}^{m} C(y)q^{y}p^{m-y}, \tag{1.4.6}$$

where $C(y)$ is the number of failure sets of size y of the module, q is the element *down* probability, and m is the number of components in the module. $C(y)$ is expressed via the module cumulative D-spectrum $F(y)$:

$$C(y) = F(y) \cdot m!/(y!(m-y)!). \tag{1.4.7}$$

It is convenient to have a formula for $p^{\star} = 1 - q^{\star}$ in a "standard" form via the number of path sets of the module. Omitting elementary calculations, see [19], p. 80, we present it as

$$p^{\star} = \sum_{x=1}^{m} K(x)p^{x}q^{m-x}, \tag{1.4.8}$$

where $K(x)$ is the number of path-sets of the module with exactly x elements *up* and the remaining $(m - x)$—*down*. $K(x)$ is expressed via the cumulative D-spectrum of the module in the following form:

$$K(x) = [1 - F(m - x)]m!/(x!(m - x)!).\qquad(1.4.9)$$

Here $x = 1, \ldots, m$, and $F(0) = 0$.

In order to obtain the spectrum of the whole network we must substitute the expressions for $q^\star$ and $p^\star = 1 - q^\star$ (1.4.6) and (1.4.8) into (1.4.9) and bring the whole expression to the homogeneous (pq)-polynomial form. This polynomial has degree $n \cdot m$ and the coefficient at $q^z(1 - p)^{(nm-z)}$ is the number $H(z)$ of failure sets of the resulting network with exactly z components *down*, $z = 1, \ldots, nm$, and $(nm - z)$ components *up*. $H(z)$ and the network cumulative D-spectrum are connected via the familiar expression

$$H(z) = F_N(z) \cdot (n \cdot m)! / (z! \cdot (n \cdot m - z)!).\qquad(1.4.10)$$

Example 1.4.2 (Series–parallel system of three bridge modules, Fig. 1.6b)
Suppose we have an organizing network S consisting of two modules 1,2 in parallel, connected in series to module 3, see Fig. 1.6b. Each module is a bridge network. It is easy to see that S has one failure set of size 1, three failure sets of size 2 and one failure set of size 3. From here it follows that its *DOWN* probability equals

$$Q_S = q^\star(p^\star)^2 + 3(q^\star)^2 p^\star + (q^\star)^3.$$

The cumulative D-spectrum of the bridge network is

$$F(1) = 0,\ F(2) = 1/5,\ F(3) = 4/5,\ F(4) = 1,\ F(5) = 1.$$

From here bridge module *down* probability $q^\star$ is

$$q^\star = 2q^2 p^3 + 8q^3 p^2 + 5q^4 p + q^5.$$

Similarly, it is easy to establish that

$$p^\star = 1 - q^\star = 2p^2 q^3 + 8p^3 q^2 + 5p^4 q + p^5.$$

Now substitute the expressions of $p^\star$ and $q^\star$ into Q_S and bring it to the (pq)-polynomial form. We omit some algebra and present below the coefficients of q^x, $x = 1, 2, \ldots, 15$:

0, 2, 28, 179, 703, 1891, 3597, 4803, 4445, 2899, 1357, 455, 105, 15, 1.

Using the formula (1.4.10), we obtain the cumulative D-spectrum, see the Table 1.1.

Spizzichino et al. [21] present an example of the D-spectrum for a bridge structure consisting of similar bridge-type modules. Spizzichino et al. [21] discusses in detail also the interaction of two combinatorial approaches to constructing D-spectra for series, parallel and recurrent networks which are based on dual interplay in considering network *UP* and *DOWN* probabilities via the system failure (cut) sets and path sets.

Table 1.1 D-spectrum of the system of three bridge modules	x	$F(x)$	x	$F(x)$	x	$F(x)$
	1	0	6	0.377822	11	0.994139
	2	0.0190476	7	0.558974	12	1
	3	0.0615385	8	0.748387	13	1
	4	0.131136	9	0.888112	14	1
	5	0.234099	10	0.965368	15	1

1.4.2 Generalized Series and Parallel Multistate Systems

We can extend the notions of parallel and series connection of components traditionally used in Reliability theory to the multi-state systems. Suppose that we have two systems S_A and S_B. The first system can be in the following states $J_A = K_A, K_A - 1, \ldots, 1, 0$, and the second—in the states $J_B = K_B, K_B - 1, \ldots, 1, 0$. When considered separately, the system A has, by definition, the failure state $DOWN_A$ when it is in the states with the number $J_A < L_A$. Similarly, the failure state for the $DOWN_B$-system is defined as the set of states with the numbers $J_B < L_B$.

Definition 1.4.1
The system $S = (S_A \wedge S_B)$ is called G-parallel connection of systems S_A and S_B if its $DOWN$ state $DOWN_S$ is defined as

$$DOWN_S = (J_A < L_A) \bigcap (J_B < L_B).$$

The system $S = (S_A \vee S_B)$ is called G-series connection of systems S_A and S_B if its $DOWN$ state $DOWN_S$ is defined as

$$DOWN_S = (J_A < L_A) \bigcup (J_B < L_B).$$

"G" stands for *generalized*.

It is worth noting that the above definition is different from the definition traditionally used in the theory of multi-state systems, see e.g. [15, 16]. Note also, that G-series and G-parallel systems can not be represented in traditional way, similar to shown on Fig. 1.6a.

Our goal is to find the D-spectrum for the G-parallel and G-series system. In fact, this problem can be easily solved along the lines of the solution for parallel and/or series connection considered in Sect. 1.4.1. For this purpose, we must determine the cumulative D-spectra for the subsystems A and B related to their failure states $DOWN_A$ and $DOWN_B$. Let us denote these spectra as $F_A(x)$, $x = 1, 2, \ldots, n_A$ and $F_B(y)$, $y = 1, 2, \ldots, n_B$, respectively. From now on act exactly as in Sect. 1.4.1.

Example 1.4.3 (G-parallel connection of two networks, see Fig. 1.7)
In this example, A is a four terminal network shown on the left part of Fig. 1.7. and B is a three terminal network shown on the right part of the figure. In both networks

Fig. 1.7 Two networks in
G-parallel connection

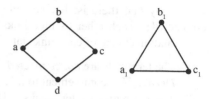

the components subject to failure are the links. $DOWN_A$ and $DOWN_B$ are defined as
the presence of two or more clusters in a network. Network A always disintegrates
into two clusters after the failure of two links. The same is true for network B. The
cumulative D-spectra are, therefore

$$F_A(1) = 0, \ F_A(2) = 1, \ F_A(3) = F_A(4) = 1; \ F_B(1) = 0,$$

$$F_B(2) = F_B(3) = 1.$$

From here it follows that

$$P(DOWN_A) = 6q^2 p^2 + 4q^3 p + q^4,$$

and

$$P(DOWN_B) = 3q^2 p + q^3.$$

By the definition of G-parallel system $S = S_A \wedge S_B$,

$$P(DOWN_S) = P(DOWN_A) \cdot P(DOWN_B) = |\text{after simple algebra}|$$
$$= 18q^4 p^3 + 18q^5 p^2 + 7q^6 p + q^7.$$

Using (1.4.3), it is easy to obtain the cumulative D-spectrum of the G-parallel system
$(n_A + n_B = 7)$:

$$F(1) = F(2) = F(3) = 0, \ F(4) = 18/35, \ F(5) = 36/42, \ F(6) = F(7) = 1.$$

1.5 Networks with Colored Links

Suppose, each link of the network N can be in r states, $r > 2$. To keep the exposition
simple we consider $r = 3$. We denote these states as 0, 1 and 2. State 0 means that
the link is erased (does not exist). For better visualization assume that state 1 means
that the link is colored *blue* and state 2—that it is colored *green*.

Link state is chosen randomly. Each link, independently of others, is in state 0
(erased) with probability p_0, in state 1 (blue) with probability p_1, and in state 2
(green) with probability $p_2 = 1 - p_0 - p_1$.

We say that there is, for example, a blue T-terminal set if the terminals are connected to each other by blue links. Let us define the following network states after random choice of each link state.

1. The terminals are *not connected* to each other. This network state is defined as DOWN. The complement to this set is denoted as UP.
2. There is a *green* T-terminal set. Denote this network state as *UPgreen*.
3. There is a *blue* T-terminal set. Denote this network state as *UPblue*.

$P(DOWN)$ is already known. Indeed, we can imagine a binary situation: either a link is erased with probability $q = p_0$ or not erased (colored in blue or green), with $p = 1 - p_0$. Then

$$P(DOWN) = \sum_{x=1}^{n} C(x) p_0^x (1 - p_0)^{n-x}, \qquad (1.5.1)$$

where $C(x)$ is expressed via the cumulative D-spectrum in a usual way

$$C(x) = F(x) \cdot n!/(x!(n - x)!).$$

To obtain $P(UPgreen)$ imagine that *all non green* edges are erased. Then we are in a binary situation and can apply (1.5.1) with an obvious change of UP to UPgreen and p to $p_2, q = 1 - p_2$:

$$P(UPgreen) = 1 - \sum_{x=1}^{n} C(x)(1 - p_2)^x p_2^{n-x}. \qquad (1.5.2)$$

Similarly, we can obtain the probability of $P(UPblue)$:

$$P(UPblue) = 1 - \sum_{x=1}^{n} C(x)(1 - p_1)^x p_1^{n-x}. \qquad (1.5.3)$$

Finally,

$$P(UPcolor) = 1 - P(DOWN).$$

Let us consider an example.

Example 1.5.1 (Complete 6-node graph with colored links, Fig. 1.8)
Below is the simulated edge D-spectrum for all-node connectivity:

$$f_5 = 0.002050, \ f_6 = 0.010155, \ f_7 = 0.029820, \ f_8 = 0.071937,$$
$$f_9 = 0.155212, \ f_{10} = 0.298834, \ f_{11} = 0.431992.$$

Suppose that $p_0 = 0.2$, $p_1 = 0.3$, $p_2 = 0.5$. The calculations give the following results:

Fig. 1.8 K_6 graph

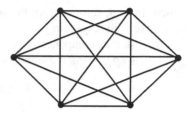

$$P(DOWN) = 0.00196122, \ P(UPgreen) = 0.814912,$$
$$P(UPblue) = 0.316939.$$

Note, that the sum of these probabilities exceeds 1 since *UPgreen* and *UPblue* have a nonempty intersection.

1.6 Network Component Importance

1.6.1 Birnbaum Measure of Component Importance

In this section, we will introduce so-called Birnbaum Importance Measure (BIM) for system components, see [1, 2]. In simple words, BIM of component j (denoted BIM_j) is the gain of system reliability obtained by replacing *down* component j by an absolutely reliable one. Formally, BIM_j is defined as follows.

Definition 1.6.1 (BIM_j).

$$BIM_j = \Psi(p_1, p_2, \ldots, 1_j, \ldots, p_k) - \Psi(p_1, p_2, \ldots, 0_j, \ldots, p_k).$$

Equivalently, BIM_j can be represented via the function

$$G(p_1, p_2, \ldots, p_k) = 1 - \Psi(p_1, p_2, \ldots, p_k)$$

as

$$BIM_j = G(p_1, p_2, \ldots, 0_j, \ldots, p_k) - G(p_1, p_2, \ldots, 1_j, \ldots, p_k), \quad (1.6.1)$$

where $G(p_1, p_2, \ldots, 0_j, \ldots, p_k)$ is the probability that our system is *DOWN* when component j is permanently *down*, and $G(p_1, p_2, \ldots, 1_j, \ldots, p_k)$ is the probability that our system is *DOWN* when component j is permanently *up*.

Among many measures of importance, Birnbaum's measure is the most popular and useful. It follows from the fact that

$$BIM_j = \frac{\partial \Psi(p_1, p_2, \ldots, p_j, \ldots, p_k)}{\partial p_j}, \quad (1.6.2)$$

which, in turn, follows from the pivotal decomposition formula (1.2.4) by taking derivative with respect to p_j.

Suppose that component j is replaced by more reliable one having reliability $p_j^* = p_j + \delta p_j$. Then, the main part of system reliability increment will be

$$\delta R = \text{BIM}_j \cdot \delta p_j.$$

The knowledge of component BIMs is the key element in finding the optimal system reinforcement strategy.

The use of BIM in reliability practice was very limited since typically the system reliability function $\Psi(\cdot)$ is not available in explicit form. It turns out that in the case of equal component reliability $p_j \equiv p$, there is a surprising connection between the BIMs and the network D-spectrum and its modification called BIM-spectrum which allows estimating and ranking the component BIMs without knowing the analytic form of system reliability function, see [10], Chap. 10.

1.6.2 BIM-Spectrum

In this section, we introduce a new combinatorial object called *BIM-spectrum*. It is closely connected to the D-spectrum (Sect. 1.3). Suppose the network has k elements subject to failure, and these elements are numbered as $1, 2, 3, \ldots, k$. Consider the set of all $k!$ permutations of these numbers. A particular permutation is denoted as $\pi = (i_1, i_2, \ldots, i_k)$. As it was described in Sect. 1.3, for each particular π we consider the destruction process of turning *down* the elements of π by moving from left to right.

Definition 1.6.2 (*BIM-spectrum*)
Let $N(x; 0_j)$ be the number of permutations satisfying the following two conditions:

(i) If the first x elements in the permutation are *down*, then the network is *DOWN*;
(ii) Element j is among the first x elements of the permutation.

The collection $\{z(x, j) = N(x; 0_j) \cdot (x!(k-x)!/k!)\}$ for a fixed j and $x = 1$, $2, \ldots, k$ is called the BIM_j-spectrum, or the *importance spectrum* of component j.

The collection of all $\{z(x; j), x = 1, 2, \ldots, k\}$ for $j = 1, \ldots, k$ is called system *BIM-spectrum* and denoted $\text{BIM} \diamond S$.

Denote by $N(x)$ the number of permutations satisfying (**i**) only. Denote by $N(x; 1_j) = N(x) - N(x; 0_j)$.

The main result of Sect. 1.6 is the following theorem proved in [10], p. 144.

Theorem 1.6.1 (Formula for BIM_j)
Let $p_i \equiv p$. *Then*

$$\text{BIM}_j = \sum_{x=1}^{k} \frac{N(x; 0_j)q^{x-1}(1-q)^{k-x} - N(x; 1_j)q^x(1-q)^{k-x-1}}{x!(k-x)!}. \qquad (1.6.3)$$

Proof Recall that

$$\text{BIM}_j = G(p_1, p_2, \ldots, 0_j, \ldots, p_k) - G(p_1, p_2, \ldots, 1_j, \ldots, p_k). \qquad (*)$$

The number of permutations such that the first x components in them, being *down*, create a *DOWN* state, and *down* component e_j is among these components, equals $N(x; 0_j)$. Each fixed permutation counted in $N(x; 0_j)$ creates a *DOWN* state which has probability $q^{x-1}(1-q)^{(k-x)}$. Taking into account that a particular system state with x components *down* and $k - x$ components *up* is repeated $x!(k-x)!$ times in different permutations, we conclude that the first term in (1.6.3) is equal to the first term in (*). Similarly, we argue that the second term in (1.6.3) equals the second term in (*), which concludes the proof.   □

The main value of this theorem is that it shows that the component BIMs are closely related to system combinatorial parameters. From the computational point of view, it is important to note that computing the values $N(x; 0_j)$ can easily be carried out by means of a minor modification of the Monte Carlo algorithm for estimating system D-spectra.

Theorem 1.6.2
Suppose that we are given the BIM $\diamond$ S for our network. Let us fix two indices α and $\beta \neq \alpha$.
If for all $x, i = 1, 2, \ldots, k$, $z(x, \alpha) \geq z(x, \beta)$, then $\text{BIM}_\alpha \geq \text{BIM}_\beta$ for all q values.

We omit the proof of this theorem, see [10] Chap. 10.

Example 1.6.1 (BIM for the edges of H_3.)
Consider the hypercube H_3 network shown on Fig. 1.9. Elements subject to failure are the edges. Nodes 1, 3, and 6 are terminals. Network failure is defined as the loss of terminal connectivity. The *UP* state is therefore the situation when all three terminals are connected to each other.

As it follows from Theorem 1.6.2, for ranking elements by their BIMs, it is sufficient to compare elements BIM-spectra. Table 1.2 presents the estimated BIM-spectra for three edges 1, 7, and 10, based on 10,000 Monte Carlo replications. It is seen from the table that $z(x; 1) \geq z(x; 7) \geq z(x; 10)$, for all x except $x = 9, 10$.

The violation of the domination of edge 1 over edge 7 for $x = 9, 10$ we explain by random error in estimating the BIM-spectra which for $M = 10,000$ replications has standard deviation 0.009.

Therefore, comparing these three edges we can conclude that edge 1 is the "most important" and edge 7 is the "second important" among the three edges. Note that comparing BIM spectra of *all* 12 edges of the given network, we arrive at the conclusion that there are three groups of edges ranked by their importance. The first group consists of equally important edges 1, 4, 5. In the second group there are edges 2, 3,

Fig. 1.9 Hypercube H_3

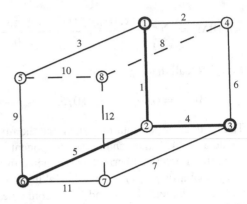

Table 1.2 Simulated BIM-spectra for nodes 1, 7, 10

x	$z(x; 1)$	$z(x; 7)$	$z(x; 10)$
2	0.0	0.0	0.0
3	0.0052	0.0039	0.0
4	0.0345	0.0277	0.0133
5	0.1298	0.1210	0.0844
6	0.3396	0.3238	0.2689
7	0.5285	0.5086	0.4769
8	0.6541	0.6416	0.6310
9	0.7515	0.7539	0.7486
10	0.8346	0.8368	0.8363
11	0.9187	0.9180	0.9176
12	1	1	1

6, 7, 9, 11, and in the third - 8, 10, 12. This ranking has clear intuitive explanation. Indeed, the edges from the first group have the common property: one node of each edge is a terminal and the second is on the distance 1 from two other terminals. In the second group, one node of each edge is some terminal and the second is on the distance 1 from only one terminal node. Three remaining edges are in the third group.

1.6.3 BIMs for Network with Several States

Obviously, the notion of BIM is applicable to any network reliability criterion. In the previous section, we considered an example when the network failure was the loss of terminal connectivity. If we imagine that in Example 1.6.1 the edge failures happen in distinct time instants, the first stage of violating terminal connectivity will be a separation of one terminal of two others, and the second (and final) stage—the isolation of all three terminals of each other.

We could therefore take more detailed approach and declare system failure as a separation of *a terminal* from two others and calculate the component importance for this failure definition. We could declare the system failure as the isolation of all three terminals and calculate another set of BIMs for network components. We see therefore that for multistate systems we can define several sets of component importance measures.

Let us describe in more detail the BIM calculation for the situation when the network has $K+1$ states, $K > 1$ (or the network has $h = K+1$ terminals). Network states are $J = K$, $J = K - 1, \ldots$, $J = 1$, $J = 0$. $J = K$ means that network is in the *UP* state (a single cluster). $J = K - 1$ means that network is separated into two isolated clusters, and so on. If $J = 0$, then network is in complete collapse, i.e. all terminals are isolated from each other, and there are $K + 1$ clusters. Denote by $P_N(J; \mathbf{p})$ the probability that the network is in state J, $J = K, K - 1, \ldots, 0$, and $\mathbf{p} = (p_1, p_2, \ldots, p_k)$.

In the multistate situation we define K BIM-spectra, for each $J = K, K-1, \ldots, 1$, in the following way. Fix some J_0, $1 \leq J_0 \leq K$. Define all states $J \geq J_0$ as the *UP* state and all other states $J < J_0$ as the *DOWN* state. Now for each J_0 we have a dichotomy of all network states into *UP* and *DOWN*, and the definition of BIM and of the BIM-spectra given in the previous section may be applied to this binary situation. Let us reformulate the BIM definition.

Definition 1.6.3
For each J_0, $1 \leq J_0 \leq K$, element's j BIM is defined as

$$\text{BIM}_j = \frac{\partial R(p_1, \ldots, p_k)}{\partial p_j} = R(p_1, \ldots, 1_j, \ldots, p_k)$$
$$- R(p_1, \ldots, 0_j, \ldots, p_k) = G(p_1, \ldots, 0_j, , \ldots, p_k)$$
$$- G(p_1, \ldots, 1_j, \ldots, p_k), \qquad\qquad (1.6.4)$$

where

$$R(\mathbf{p}) = \sum_{J_0 \leq J \leq K} P_N(J; \mathbf{p}),$$

and the meaning of $R(p_1, \ldots, 1_j, \ldots, p_k)$ and $R(p_1, \ldots, 0_j, , \ldots, p_k)$ is as described in the BIM definition in the previous section.

Now for each J_0, $1 \leq J_0 \leq K$, the BIM-spectrum definition is the same as in the previous section, because we reduced the multistate case to the binary *UP, DOWN* case. We remind that we consider the case of $p_i \equiv p$.

Example 1.6.2
Let us return to the network shown on Fig. 1.9. Now we have three states corresponding to $J_0 = 2$, $J_0 = 1$ and there are therefore two BIM-spectra. The first (for $J_0 = 2$) has been already presented in the Table 1.2. Table 1.3 presents the estimated BIM-spectra for the same three edges 1, 7, and 10, for $J_0 = 1$. It is seen from the table that for each x, $z(x; 1) \geq z(x, 7) \geq z(x, 10)$. We see therefore that these elements are ranked exactly as in the previous *UP–DOWN* case. The data (not presented here) say that BIM ordering is the same as in the previous example. Our conjecture (so far

Table 1.3 Second simulated
BIM-spectra for nodes
1, 7, 10

x	$z(x; 1)$	$z(x; 7)$	$z(x; 10)$
6	0.0123	0.0096	0.0058
7	0.1268	0.0968	0.0871
8	0.3009	0.2565	0.2508
9	0.4958	0.4532	0.4523
10	0.6723	0.6511	0.6501
11	0.8347	0.8316	0.8287
12	1	1	1

based only on limited experimental data) is that BIM ranking for multistate network is very little influenced by the choice of J_0.

1.6.4 Joint Reliability Importance

Joint Reliability Importance (JRI) for two components have been introduced by Hong and Lie [13] as a measure of components interaction in determining system reliability, see also [23]. JRI for components i and j is defined as

$$JRI_{(i,j)} = \frac{\partial^2 \Psi(\mathbf{p})}{\partial p_i \partial p_j}.$$

Before we proceed, let us analyze the role of second derivatives in system reliability increase if two components i and j simultaneously get reinforced and their initial reliability p_i and p_j is increased by Δp_i and Δp_j, respectively. It is easy to establish, using pivotal decomposition with respect to components i and j, that the second derivatives of type

$$\frac{\partial^2 \Psi(p_1, \ldots, p_k)}{\partial p_i^2}$$

are equal zero. Then the Taylor series expansion gives that

$$\Psi(p_1, \ldots, p_i + \Delta p_i, \ldots, p_j + \Delta p_j, \ldots, p_k) \approx \Psi(p_1, \ldots, p_i, \ldots, p_j, \ldots, p_k)$$
$$+ \frac{\partial \Psi(p_1, \ldots, p_k)}{\partial p_i} \Delta p_i + \frac{\partial \Psi(p_1, \ldots, p_k)}{\partial p_j} \Delta p_j + \frac{\partial^2 \Psi(p_1, \ldots, p_k)}{\partial p_i \partial p_j} \Delta p_i \Delta p_j.$$

If we take into account that $|\partial^2 \Psi(p_1, \ldots, p_k)/\partial p_i \partial p_j| \leq 1$, (which is easy to prove), we conclude that the JRI's impact on more accurate reliability estimation is rather small relative to the impact of the BIMs.

We present Theorem 1.6.3 for computing the JRI. It is based on using pivotal formula and combinatorial arguments and resembles the derivation of BIM_j, see [20] for the detailed proof.

Let $\delta_i = 1/0$ be the indicator variable for the component i, which is equal 0 if i is among the x components turned *down* and is equal 1, otherwise. Similarly we define δ_j for component $j \neq i$. Let $F(x; \delta_i, \delta_j)$ be the joint probability of the events $(Y \leq x)$, δ_i and δ_j. For example, $F(x; 0_i, 0_j)$ is the probability that the system is *DOWN* after x components are turned *down* in permutation π, *and* components i *and* j are among them.

Theorem 1.6.3

$$
\begin{aligned}
\mathrm{JRI}_{(ij)} = k! \Bigg[& \sum_{x=1}^{k} F(x; 1_i, 0_j) q^{x-1} p^{(k-x-1)} / (x!(k-x)!) \\
& + \sum_{x=1}^{k} F(x; 0_i, 1_j) q^{x-1} p^{(k-x-1)} / (x!(k-x)!) \\
& - \sum_{x=1}^{k} F(x; 0_i, 0_j) q^{x-2} p^{(k-x)} / (x!(k-x)!) \\
& - \sum_{x=1}^{k} F(x; 1_i, 1_j) q^{x} p^{(k-x-2)} / (x!(k-x)!) \Bigg]
\end{aligned}
\tag{1.6.5}
$$

1.7 Reliability Gradient

1.7.1 Border States. Evolution Process

In this section, we will consider a network with components having different *up* probabilities: $P(\text{component } i \text{ is up}) = p_i$, $i = 1, \ldots, k$. In this model, the technique developed in the previous section does not work and we have to develop new approach to finding the reliability gradient vector.

Suppose that we have a binary network with two states, *UP* and *DOWN*. The probability that the network is *UP* is given by the following function

$$
R_N = \Psi(p_1, p_2, \ldots, p_k).
$$

Our main goal is to develop a method for estimating the so-called *reliability gradient* vector.

Definition 1.7.1 *Reliability gradient vector.*
The reliability gradient vector ∇R_N is the vector whose components are the partial derivatives of $\Psi(\cdot)$:

$$
\nabla R_N = \left(\frac{\partial \Psi}{\partial p_1}, \ldots, \frac{\partial \Psi}{\partial p_k} \right).
\tag{1.7.1}
$$

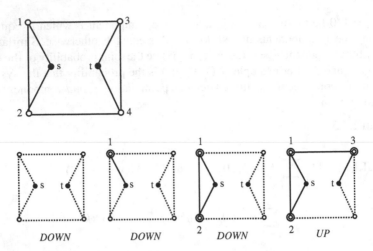

Fig. 1.10 The network with two terminals (*bold*) and its evolution process. Nodes are born in the following order: $1 \rightarrow 2 \rightarrow 3$ (*double circled*)

A computationally efficient algorithm for Monte Carlo estimation of ∇R_N is based on identifying so-called *border states* of the network, see [10], Chap. 5 and [7]. Before we give their definition, let us consider an example.

Example 1.7.1 (Evolution of a small s–t network with 4 nonterminal nodes) Figure 1.10 shows a trajectory of the random process of gradual *node* births until the network becomes *s–t* connected. When a node is born, all links incident to this node become "alive".

When node 3 is born, the network becomes *s–t* connected.

The state denoted by $DN^{\star}$ directly preceding the *UP* state is the so-called *border* state. Formally, we give the following

Definition 1.7.2 (*Border state*)
Border state is a network *DOWN* state whose Manhattan distance from the *UP* state is 1.

This definition says that a vector $\mathbf{v} = (x_1, x_2, \ldots, x_k)$ with binary components determines a border state if

(i) it is a *DOWN* state, i.e. $\varphi(\mathbf{v}) = 0$ and
(ii) there is a vector $\mathbf{w} = (0, 0, \ldots, 0, 1, 0, \ldots, 0)$ such that $\varphi(\mathbf{v} + \mathbf{w}) = 1$.

In the example above, $\mathbf{v} = (1, 1, 0, 0)$ and $\mathbf{w} = (0, 0, 1, 0)$.

The next step is to identify the set $\Gamma(\mathbf{v})$ of all components whose addition to the border state $\mathbf{v} \in DN^{\star}$ transfers the network into the *UP* state. So, for example, for the border state shown on Fig. 1.10, adding the nodes 3 or 4 transfers the network into the *UP* state. Thus, here for $\mathbf{v} = (1, 1, 0, 0)$, $\Gamma(\mathbf{v}) = \{3, 4\}$. We will call $\Gamma(\mathbf{v})$ the *activating set* of the border state $\mathbf{v}$.

It is important to note that the border states and their activating sets are topological parameters of the network and of the definition of its *DOWN* state. Similarly to the D-spectra, they do not depend on the probabilistic mechanism governing network components failures.

To proceed further we will introduce an artificial *evolution process* on network components. Assume that at time $t = 0$ all components are *down*, and component i, independently of others, is born at the random instant ξ_i, which has Exponential distribution with parameter λ_i. Let us remind that it means the following:

$$P(\xi_i \leq t) = p_i(t) = 1 - \exp^{-\lambda_i t}, t > 0.$$

Before the birth, the component is *down*. Once it is born, it becomes *up* forever. Thus, $p_i(t_0)$ is the probability that the component was born on $[0, t_0]$ and is *up* at the instant t_0. The complementary probability $q_i(t_0) = 1 - p_i(t_0)$ is the probability that at t_0 the component i is *down* (has yet not born).

For a fixed time instant $t = t_0$, we will denote $p_i(t_0) = p_i$ and $1 - p_i(t_0) = q_i$.

Now we remind to the reader the fundamental property of the Exponential distribution (see e.g. [1, 8, 10]) that a component i which was not born before the instant t_0, will be born in the small interval $[t_0, t_0 + \delta t]$ with probability $\lambda_i \cdot \delta t + o(\delta t)$, $\delta t \to 0$ *not depending* on t. (This is the so-called *memoryless* property). λ_i is called the birth rate of component i.

Let us denote by $\Lambda(\mathbf{v})$ the sum of all birth rates of the set $\Gamma(\mathbf{v})$ which activates the border state $\mathbf{v}$:

$$\Lambda(\mathbf{v}) = \sum_{j \in \Gamma(\mathbf{v})} \lambda_j.$$

Let us call $\Lambda(\mathbf{v})$ the *flow* from the state $\mathbf{v}$. Let $P(\mathbf{v}; t)$ be the probability that the network is in state $\mathbf{v}$ at time t.

1.7.2 Gradient Formula

Let us denote by $R_N(t) = \Psi(p_1(t), p_2(t), \ldots, p_k(t))$ the probability that the network is *UP* at the instant t. We will derive a differential equation for $R_N(t)$. Below is the standard reasoning leading to this differential equation borrowed from [7, 10].

Let us consider the event "the network is *UP* at the instant $t + \delta t$". This event takes place if and only if the network was already in *UP* at the instant t or it was at time t in one of its border states and went into *UP* in the interval $[t, t + \delta t]$. All other possibilities which involve more than one transition during this interval are of magnitude $o(\delta t)$ when $\delta t \to 0$. This leads to the following relationship between $R_N(t + \delta t)$ and $R_N(t)$:

$$R_N(t + \delta t) = R_N(t) + \sum_{v \in DN^*} P(\mathbf{v}; t)(v)\delta t + o(\delta t).$$

Now transfer $R_N(t)$ to the left-hand side, divide both sides by δt and set $\delta t \to 0$. We arrive at the following relationship:

$$\frac{dR_N(t)}{dt} = \sum_{v \in DN^*} P(\mathbf{v}; t)\Lambda(v). \tag{1.7.2}$$

Now recall that $R_N(t) = R_N(p_1(t), p_2(t), \ldots, p_k(t))$ and represent the left-hand side of (1.7.2) using the chain rule of differentiation:

$$\frac{dR_N(p_1(t), p_2(t), \ldots, p_k(t))}{dt} = \sum_{j=1}^{k} \frac{\partial R_N}{\partial p_j} \cdot \frac{dp_j(t)}{dt}. \tag{1.7.3}$$

Now take into account that

$$p_j(t) = 1 - e^{-\lambda_j t}, \quad \frac{dp_j(t)}{dt} = \lambda_j e^{-\lambda_j t} = \lambda_j q_j(t),$$

and rewrite the left-hand side of (1.7.2) in the following form:

$$\frac{dR_N(t)}{dt} = \sum_{j=1}^{k} \frac{\partial R_N}{\partial p_j} \cdot q_j(t) \cdot \lambda_j = \nabla R_N \bullet \{q_1(t)\lambda_1, q_2(t)\lambda_2, \ldots, q_k(t)\lambda_k\}. \tag{1.7.4}$$

The sign " $\bullet$ " is a shorthand notation for the scalar product of vectors $\nabla R_N = \{\partial R_N/\partial p_1, \partial R_N/\partial p_2, \ldots, \partial R_N/\partial p_k\}$ and $\{q_1(t)\lambda_1, q_2(t)\lambda_2, \ldots, q_k(t)\lambda_k\}$.

Comparing (1.7.2) and (1.7.4) and setting $t = t_0$ we arrive at the desired relationship which expresses the gradient vector via the probabilities of the border states and the corresponding flows.

Theorem 1.7.1

$$\nabla R_N \bullet \{q_1\lambda_1, q_2\lambda_2, \ldots, q_k\lambda_k\} = \sum_{v \in DN^*} P(\mathbf{v}; t)\Lambda(v). \tag{1.7.5}$$

To apply (1.7.5) to finding $\partial R_N/\partial p_i$, we have to regroup the terms in the right-hand side and to find the probability of all border states which become "activated" by component i, i.e. whose probabilities are multiplied by λ_i. Let us show how it works on an example of a small s–t network.

Example 1.7.2 (s–t network with unreliable nodes, Fig. 1.10)
Let us find out all network *UP* states. There are seven such states:

$$(1, 1, 1, 1), (1, 1, 1, 0), (1, 1, 0, 1), (1, 0, 1, 1), (0, 1, 1, 1); (1, 0, 1, 0), (0, 1, 0, 1).$$

Node i is *up* with probability p_i. Denote by $H = \prod_{i=1}^{4} p_i$. It is easy to find that

$$R_N = H \cdot [1 + (1 - p_2)/p_2 + (1 - p_3)/p_3 + (1 - p_4)/p_4 \\ + (1 - p_2)(1 - p_4)/p_2 p_4 + (1 - p_1)(1 - p_3)/p_1 p_3 + (1 - p_1)/p_1].$$

From here it follows after some algebra that

$$\partial R_N / \partial p_1 = (1 - p_2) p_3 p_4 + p_2 (1 - p_4) p_3 + (1 - p_2)(1 - p_4) p_3.$$

There are four border states with two adjacent nodes in *up*:

$$(1, 1, 0, 0),\ (1, 0, 0, 1),\ (0, 0, 1, 1),\ (0, 1, 1, 0),$$

and four border states with only one node in *up*:

$$(1, 0, 0, 0),\ (0, 1, 0, 0),\ (0, 0, 1, 0),\ (0, 0, 0, 1).$$

Only three of the above border states are activated by the birth of node 1:

$$(0, 0, 1, 1),\ (0, 1, 1, 0),\ (0, 0, 0, 1).$$

The probability of these states is equal to

$$(1 - p_1) \cdot [(1 - p_2) p_3 p_4 + p_2 (1 - p_4) p_3 + (1 - p_2)(1 - p_4) p_3].$$

The multiple at $(1 - p_1) = q_1$ equals to $\partial R_N / \partial p_1$, in accord with Theorem 1.7.1.

1.8 Basic Monte Carlo Algorithms

This section presents a non formal description of three principal algorithms which serve as a basis for reliability calculations in this book. A detailed description of these algorithms and their properties can be found in [10].

1.8.1 Testing the Network Terminal Connectivity

Checking network terminal connectivity is a common task for considerable part of network algorithms. Among many methods available for this task, the following method is very convenient for all algorithms considered in this book (and also [10]). It is based on so called *Disjoint Set Structures* (DSS), see [10].

Suppose that $\mathbf{N}' = (V', E', T)$ is some state of the network $\mathbf{N} = (V, E, T)$, i.e $E' \subseteq E$, and V' is the set of nodes incident to edges from E'. It is necessary to check whether each pair of terminals from T is connected by some path consisting of edges from E'.

The idea of the method is as follows.

1. Let us initially associate each node $v \in V'$ with a component C_v consisting of only one node v.
2. Suppose that is given some permutation π of edges from E'.

3. Choose sequentially the edges from π moving from left to right.Suppose $e \in E'$, $e = (a, b)$ is chosen. If the nodes a and b belong to *different* components, *merge* these components. From now on the new component will contain both nodes a and b and the 'old' components will not exist anymore. If the nodes a and b belong to the *same* component do nothing. Note that initially each component consists of exactly one node but after some steps this is not so. Note also that if two nodes belong to one component, then these nodes are connected by the edges from E'.

4. After all edges were chosen, it must be checked whether there exists a component which contains all terminals. If yes, then $\mathbf{N'}$ is T-terminal connected.

1.8.2 Estimating the D-Spectra and Component Importance

By definition, D-spectra and BIM-spectra are closely connected to analyzing permutations of network components and fixing for each permutation the position of the component which, being turned *down*, causes the appearance of certain event. For example, this event may be network passage from state $K - I$ into the state $K - I - 1$ (see Sect. 1.3.2), or entering the state when the size of maximal isolated cluster becomes less than some fixed number (see *Remark* in Sect. 1.3.5).

Suppose that number of network elements equals n.

For each type of D-spectrum, and for each $i = 1, \ldots, n$, we must *estimate* the number of permutations satisfying an appropriate condition at the moment i. Afterwards, we calculate the estimates of the D-spectrum. For example, we see that substituting into (1.3.1) the estimated numbers of permutations having $r_i(\pi) = \alpha_i$ we arrive at the estimate of the multidimensional D-spectrum.

The method for estimating the number of permutations for D-spectra or BIM-spectra in general form works as follows.

1. Suppose that the criterion (*event*) is given.
   Denote by $N(D_i)$ and $N(\text{BIM}_{i,j})$, $i, j = 1, \ldots, n$ the number of permutations in M trials for D-spectrum and BIM-spectrum, respectively.
   Put $N(D_i) = 0$ and $N(\text{BIM}_{i,j}) = 0$ for all $i, j = 1, \ldots, n$.

2. *Simulate* random permutation π of network elements. Turn *down* elements of π, moving along it from the left to right. *Stop* at the index i when the defined *event* takes place.
   Note that for all kinds of spectra, for fixing the *event* we use the DSS-method, which gives full information about the network state, i.e the number of components, the size of each component, presence of terminals in components and so on.

3. For D-spectrum let $N(D_i) : = N(D_i) + 1$

4. For BIM-spectrum, check whether the following two conditions are satisfied:

   (a) the *event* takes place at the index i;
   (b) the element j is to the left from or at the place of the index i.

If (a) and (b) take place, put $N(\text{BIM}_{i,j}) := N(\text{BIM}_{i,j}) + 1$. Repeat the steps 2–4 M times.

5. The estimated number of permutations for D-spectrum and B-spectrum are equal $N(D_i) \cdot n!/M$ and $N(\text{BIM}_{i,j}) \cdot n!/M$, respectively.

1.8.3 Estimating the Gradient Vector

Let us remind that the definition of gradient vector is connected to an artificial evolution process on network elements, see Sect. 1.7.1. Initially, at $t = 0$ all elements are *down*. At some random moment, element e is "born" and remains *up* forever. This random moment is exponentially distributed with parameter $\lambda(e)$.

Fix some instant t_0 and choose $\lambda(e)$ so that $exp(-\lambda(e)t_0) = q(e)$, the *down* probability of element e.

The method used for estimating the gradient vector is based on a special graph-theoretic construction called Lomonosov's Turnip, see Chap. 9 in [10]. The idea of this method is given below.

1. Denote by $\widehat{\partial R/\partial p_i}$ the estimate of $\partial R/\partial p_i$. Put $\widehat{\partial R/\partial p_i} := 0$ for all $i, i = 1, \ldots, n$.
2. Simulate random permutation π of network elements using *transition probabilities* (for details see [10], Chap. 9).
3. Turn elements from *down* to *up* moving along π, from the left to right.
4. For each network element i do the following.

   (a) Fix the first *border* state B which may be transferred into UP state by activating element i.
   (b) Fix the first UP state U defined by the permutation π.
   (c) Denote by $\xi(B)$ and $\xi(U)$ the random moments at which the evolution process enters the states B and U, respectively. Calculate $\text{Conv} = P(\xi(B) \le t_0) - P(\xi(U) \le t_0)$ and put

   $$\widehat{\partial R/\partial p_i} := \widehat{\partial R/\partial p_i} + \text{Conv}.$$

   (The methods for calculating convolutions are described in [10], see Chap. 7 and Appendix B)

5. Repeat 2–4 M times.
6. For each i, put $\widehat{\partial R/\partial p_i}/M \cdot q_i$.

1.8.4 Accuracy of Monte Carlo Reliability Estimation

We estimate the ith element of the marginal D-spectrum $\{f_i\}$ by the ratio $\hat{f}_i = M_i/M$, where M_i is the number of cases, out of M, where the network failure was observed on the ith position of the permutation in the process of turning *down* the network components. It is known from statistics that $\hat{f}_i$ is an unbiased estimator of f_i and its variance and standard deviation are equal

$$\sigma^2 = \mathrm{Var}[\hat{f}_i] = \frac{f_i(1-f_i)}{M} \tag{1.8.1}$$

and

$$\sigma = \sqrt{\frac{f_i(1-f_i)}{M}},$$

respectively. From here it follows that the relative error *r.e.* of $\hat{f}_i$ equals

$$r.e. = \sqrt{\frac{(1-f_i)}{M \cdot f_i}}. \tag{1.8.2}$$

The *r.e.* does not exceed $1/\sqrt{M \cdot f_i}$. Thus, for example, if we want to guarantee *r.e.* < 0.1, we have to take

$$M > 100/f_i.$$

So, to estimate with *r.e.* 0.1 (10%) the probability of 0.0005. we must take $M > 100/0.0005 = 200,000$. What follows from that simple arithmetics is that in order to estimate with reasonable accuracy the first terms in the spectra (which are as a rule, the smallest ones), the number of replications must be very large. So, for accurate estimation of f_5 in the D-spectrum of 5-dimensional cube network (see e.g. [10], p. 193), the number of replications M must be of magnitude 10^8. (f_5, the first element of the spectrum is about 10^{-6}). It is important to know the first nonzero element of the D-spectrum in the case when we want to estimate the number $C_{\min}$ of min-size min cuts. It should be noted that this itself is a NP-hard problem, and there is no easy solution for it, except for network with regular structure, like cubes, butterflies, rectangular grid etc. For example, in the five-dimensional cube, the number $C_{\min} = 32$, the number of nodes in the network.

Accurate estimation of the minimal nonzero elements of the spectrum is the most critical situation and is, in fact, an exceptional and most difficult case of using Monte Carlo for approximating the D-spectra. A typical application of the D-spectra is calculating network *DOWN* probability using the cumulative D-spectrum, see for example, formula (1.8.3).

The crucial observation is that the less accurately estimated terms of the D-spectra are the smallest and their contribution to $P(DOWN)$ is rather small. On the other hand, the sum in the formula for the final result contains a large number of terms, and it is not clear how the errors in estimating each one of them may influence the final result. One way to estimate the error in calculating $P(DOWN)$ is to use the explicit formula for its variance, see e.g. [9] and [10], p. 110. Here we will present a version of bootstrap estimation of the error in reliability calculations based on Monte Carlo estimated D-spectrum.

Let us present a typical formula for computing $P(DOWN; q)$ in the following form:

$$P(DOWN; q) = \sum_{x=1}^{k} \left(\sum_{i=1}^{x} f_i \right) k! q^x (1-q)^{k-x} / (x!(k-x)!). \qquad (1.8.3)$$

Now let us simulate a random error in computing each f_i. We will do it for a typical case of $M = 10^6$ replications for estimating the spectrum. We replace f_i by

$$f_i(\varepsilon) = f_i + \varepsilon$$

where

$$\varepsilon \sim \text{Normal}(0, 1) \cdot 0.001 \cdot \sqrt{f_i \cdot (1 - f_i)}$$

is a normally distributed random variable with zero mean value and standard deviation $0.001 \cdot \sqrt{f_i \cdot (1 - f_i)}$.

Then put

$$P(DOWN; q, \varepsilon) = \sum_{x=1}^{k} \left(\sum_{i=1}^{x} f_i(\varepsilon) \right) k! q^x (1-q)^{k-x} / (x!(k-x)!). \qquad (1.8.4)$$

For a fixed value of q, compute ten replicas of $P(DOWN; q, \varepsilon)$ and calculate the *maximal* value of the observed difference

$$\delta = |P(DOWN; q) - P(DOWN; q, \varepsilon)|.$$

Since δ may change for different q, we repeat this process for several q values in order to cover a wide range of network *DOWN* probabilities, from highly reliable networks to networks with $P(DOWN; q) > 0.3$.

Table 1.4 presents the results of typical error calculation for a 5-dimensional cube network H_5. It has 32 nodes and 80 edges, edges are subject to failure. Figure 1.11 presents the plot of its D-spectrum.

We see from Table 1.4 that the δ is very small and practically does not affect the result, and we can conclude that $M = 10^6$ Monte Carlo replications provide satisfactory accuracy of the Monte Carlo estimation, far beyond the statistical errors arising in estimating the q values from empirical or expert data.

To be on the safe side, we assume that for the reliability simulation of networks of size H_5, the maximal error does not exceed ± 0.0007. If we reduce the number of replication 10 times and take $M = 10^5$, the maximal error increases by factor of ≈ 3, and is of order ± 0.002, which by our opinion, is satisfactory for most of reliability calculations.

Table 1.4 Simulated maximal errors in estimating $P(DOWN; q)$

q	$P(DOWN; q)$	max δ
0.1	0.000386	0.000025
0.2	0.010452	0.0001
0.3	0.077612	0.0006
0.4	0.296269	0.0007

Fig. 1.11 The $\{f_i\}$ spectrum for H_5 (nonreliable edges, all-node connectivity)

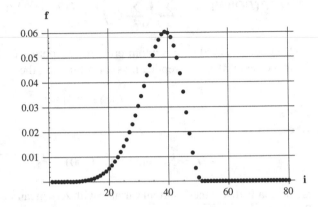

1.9 D-Spectra for Multi-Step Systems

As the reader remembers, in single-step network, single component failure either leaves the state of the network unchanged, or causes its drop by one unit. So is the situation with link failures. Suppose that the network has $K + 1$ states, denoted as $J = K$, $J = K - 1, \ldots$, $J = 0$. Thus, moving along the permutation $\pi = (i_1, i_2, \ldots, i_k)$ of link numbers and turning them from *up* to *down*, we necessarily will meet K components which, being turned *down* from *up*, caused the K transitions $K \rightarrow (K-1), (K-1) \rightarrow (K-2), \ldots, 1 \rightarrow 0$. Thus, each permutation produces one realization of each of the K anchors, each of which is the position of the component causing a transition. This leads to the fact that the marginal D-spectra $\mathbf{f}^{(i)}, i = 1, \ldots, K$ (see Definition 1.3.4) are proper discrete distributions.

This situation, however, may change if the network is not a single-step, i.e. elimination (failure) of a single component may cause the network state drop by more than one unit. Let us consider an example of a network in which the components subject to failure are the *nodes*, see Fig. 1.12. We remind that node failure means elimination of all edges incident to this node but the node remains intact.

Example 1.9.1 (Number of clusters for node failures)
The upper panel of the figure shows what happens when the nodes 1, 2, 3 fail in the order given by $\pi = (1, 2, 3)$. The number of clusters change gradually increasing by one unit after each failure: $1 \rightarrow 2 \rightarrow 3 \rightarrow 4$. Correspondingly, this permutation produces the positions 1, 2 and 3 for the first, second and third anchor, respectively.

The lower panel shows network disintegration when the nodes fail in the order given by $\pi = (3, 1, 2)$. Here the number of clusters change in the following order:

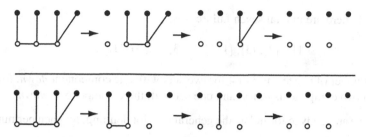

Fig. 1.12 Network with seven nodes, four terminals (*bold*)

$1 \rightarrow 3 \rightarrow 4 \rightarrow 4$: after node 3 fails, the number of clusters jumps up by *two* units. So, if we follow the rules for single-jump networks, we locate only *two* jumps and conclude that the first anchor for the transition into 2 clusters is missing, the second equals 1 and the third equals 2. In that case we will not obtain proper distributions of the marginal spectra.

The conclusion from this example is that we must change the anchor definition for multi-jump systems. So, if a node failure causes the increase of clusters from 1 to 3, we must treat this event as *two* events happened simultaneously: the change from 1 to 2, and the change from 2 to 3. As a result, we define the first anchor being equal to the second, i.e. $r_1 = r_2 = 1$. The situation for the permutations $\pi = (3, 2, 1)$ and $\pi = (2, 3, 1)$ will be the same. For $\pi = (2, 1, 3)$ we have $r_1 = r_2 = 1, r_3 = 3$; for $\pi = (1, 3, 2)$ we obtain $r_1 = 1, r_2 = r_3 = 2$. Finally, $\pi = (1, 2, 3)$ is the only permutation where the anchors are separated: $r_1 = 1, r_2 = 2, r_3 = 3$.

It is easy to obtain now the cumulative marginal spectra:

$$F^{(1)}(1) = 1; \; F^{(2)}(1) = 2/3, \; F^{(2)}(2) = 1;$$

$$F^{(3)}(1) = 0, \; F^{(3)}(2) = 2/3, \; F^{(3)}(3) = 1.$$

Here $F^{(1)}(x)$ is the probability that after x failures, the number of clusters is ≥ 2, $F^{(2)}(x)$ is the probability that after x failures, the number of clusters is ≥ 3, $F^{(3)}(x)$ is the probability that after x failures, the number of clusters is 4.

Remember that the number of failure sets of size x is connected to the D-spectrum via the formula $C(x) = F(x) \cdot k!/(x!(k-x)!)$. Since now we have *several* cumulative D-spectra, we will use the notation

$$C^{(s)}(x) = F^{(s)}(x) \cdot k!/(x!(k-x)!),$$

where

$C^{(s)}(x) = $ (number of failure sets with x components *down* and number of clusters $> s$).

It is easy to find out that $C^{(2)}(1) = 2, C^{(2)}(2) = 3, C^{(2)}(3) = 1$.

Indeed, there are in total seven failure sets:

$$\{1\}, \{2\}, \{3\}, \{1, 2\}, \{1, 3\}, \{2, 3\}, \{1, 2, 3\}.$$

The number of clusters is 3 or 4 for two sets with one component *down*, for three sets with two components *down*, and one set with three components *down*.

Now we are ready to formulate the general rule for defining, for each permutation, all its K anchors.

If at the failure of the component on ith position in π the system jumps from the state $J = K - A$ into state $J = K - A - B$, and $B > 1$, we assume that

$$r_{(A+1)}(\pi) = \cdots = r_{(A+B)}(\pi) = i.$$

Remark

Since the *marginal* cumulative D-spectra is our main tool for investigating networks in the process of their disintegration into isolated clusters, we might take an alternative approach to the multi-step networks. We simply illustrate it on the same example with a seven node network shown on Fig. 1.12.

Example 1.9.1-continued

Suppose that we accept a "dichotomic" approach and consider only two states: *UP* when there is one cluster and its complement *DOWN* (2, 3, or 4 clusters). Then, obviously, the transition $UP \rightarrow DOWN$ takes place after a single node fails, i.e. the first corresponding cumulative spectrum is

$$F^{(1)}(1) = F^{(1)}(2) = F^{(1)}(3) = 1. (*)$$

Now define the *UP* state if the number of clusters is 1 or 2 and *DOWN*—as its complement (3 or 4 clusters.) Then it is easy to see that in four permutations where the first position is occupied by nodes 2 or 3, the $UP \rightarrow DOWN$ transition takes place when the nodes 2 or 3 fail, and in the remaining two permutations, when two nodes fail. This will bring us the following cumulative D-spectrum:

$$F^{(2)}(1) = 4/6, \; F^{(2)}(2) = F^{(2)}(3) = 1. (**)$$

Finally, if *DOWN* is defined when there are 4 clusters and its complement (1, 2 or 3 clusters) is *UP*, then analyzing all six permutations we find out that the third cumulative D-spectrum will be

$$F^{(3)}(1) = 0, \; F^{(3)}(2) = 4/6, \; F^{(3)}(3) = 1. (* * *)$$

We see that $(*)$, $(**)$ and $(* * *)$ coincide with the previously obtained results.

Now we are ready to formulate the general rule for defining, for each permutation, all its K anchors.

Rule: multidimensional D − spectra for multi − step networks.

If at the failure of the component on ith position in π the system jumps from the state $J = K - A$ into state $J = K - A - B$, and $B > 1$, put

$$r_{(A+1)}(\pi) = \cdots = r_{(A+B)}(\pi) = i. \tag{1.9.1}$$

References

1. Barlow, R.E., and Proschan, F. 1975. *Statistical theory of reliability and life testing*. NY: Holt, Rinehart and Winston, Inc.
2. Birnbaum, Z.W. 1969. On the importance of different components in multicomponent system. *Multivariate Analysis-II*, ed. P.R. Krishnaiah, 581–592. New York: Academic Press.
3. Brandes, U., and T. Erlebach (eds.) 2005. *Network analysis—Methodological foundations*. Berlin: Springer-Verlag.
4. Burtin, Y., and B.G. Pittel. 1972. Asymptotic estimates of the reliability of complex systems. *Engineering Cybernetics* 10(3):445–451.
5. David, H.A. 1981. *Order Statistics*, second edition. NY: Wiley.
6. Elperin, T., Gertsbakh, I., and M. Lomonosov. 1991. Estimation of network reliability using graph evolution models. *IEEE Transactions on Reliability* R-40:572–581.
7. Elperin, T., Getsbakh, I., and M. Lomonosov. 1992. An evolution model for Monte Carlo estimation of equilibrium network renewal parameters. *Probability in Engineering and Informational Sciences* 6:457–469.
8. Gertsbakh, I.B. 1989. *Statistical reliability theory*. NY:Marcel Dekker, Inc.
9. Gertsbakh, I., and Y. Shpungin. 2004. Combinatorial approaches to Monte Carlo estimation of network lifetime distribution. *Applied Stochastic Models in Business and Industry* 20:49–57.
10. Gertsbakh, Ilya and Yoseph Shpungin. 2009. *Models of network reliability: Analysis, combinatorics, and Monte Carlo*. Boca Raton: CRC Press.
11. Gertsbakh, I., and Y. Shpungin. Stochastic models of network survivability. 2012. *Quality Technology and Quantitative Management*, Special Issue devoted to S. Zacks 9(1) to appear.
12. Gertsbakh, Ilya and Yoseph Shpungin. 2011. Multidimensional spectra of multi state systems with binary components. In *Recent Advances in Reliability: Signatures, Multi-State Systems and Statistical Inference*, eds. Frenkel, I. and A. Lisniansky, Chap. 4. Heidelberg: Springer, in press.
13. Hong, Jong, Silk, and Chang Hou Lie. 1993. Joint reliability importance of two edges in undirected network. *IEEE Transactions on Reliability* 42(1):17–23.
14. Levitin, G., Gertsbakh, I., and Y. Shpungin. 2010. Evaluating the damage associated with intentional network disintegration. *Reliability Engineering and System Safety* 96(4):433–439.
15. Lisniansky, Anatoly and Gregory Levitin. 2003. *Multi-state system reliability*. NJ: World Scientific.
16. Lisniansky, A., Ilia, Frenkel, and Y. Ding. 2010. *Multi-state reliability analysis and optimization for engineers and industrial managers*. London: Springer.
17. Navarro, J., and F. Spizzichino. 2011. Different definitions of the concept of signature and relevant properties of coherent systems. In *Recent Advances in Reliability: Signatures, Multi-State Systems and Statistical Inference*, eds. Frenkel, I. and A. Lisniansky, Chap. 3. Heidelberg: Springer, in press.
18. Samaniego, F.J. 1985. On closure of the IFR under formation of coherent systems. *IEEE Transactions on Reliability* 34:69–72.
19. Samaniego F.J. 2007. *System signatures and their applications in engineering reliability*. NY: Springer.

20. Shpungin, Y., and I. Gertsbakh. 2012. Combinatorial approach to computing importance indices in coherent systems. *Probability in Engineering and Informational Sciences* 26(1) (to appear).
21. Spizzichino, F., Shpungin, Y., and I. Gertsbakh. 2011. Signatures of coherent systems built with separate modules. *Journal of Applied Probability* 48(3).
22. Wolfram, Stephen. 1992. *MATHEMATICA: A system for Doing Mathematics by Computer*. Addison-Wesley Publishing Company.
23. Xueli, G., Liring, C., and J. Li. 2007. Analysis for joint importance of components in a coherent system. *European Journal of Operational Research* 182:282–299.

Chapter 2
Applications

Abstract In Sect. 1 we compare the probabilistic behavior (the resilience) of three networks under random attack on their nodes: (1) five-dimensional cube network; (2) a network obtained by preferential attachment; (3) a slightly modified 9/11 terrorist network. All three networks have the same number of nodes and links, and their failure is defined as disintegration into isolated components of maximal size of 10 nodes. Section 2 describes various approaches to network reliability design based on improving their reliability by means of reinforcing several nodes or several links, combined with deletion of the least important components. Section 3 analyzes various scenarios of predisaster management of a transportation network with 34 edges, 25 nodes and 4 terminals. The management is carried out by the "best" choice of the subset of links which are reinforced to provide given level of terminal connectivity, subject to budgetary constraint. Section 4 presents a probabilistic follow-up of a four-state network disintegration process when the network links fail in random order.

Keywords Nodes failures · Optimal network design · Predisaster management · Multistate network disintegration

2.1 Network Under Attack: Symmetric Versus Scale-Free Network

In this section we will compare the nodal resilience of three networks with the same number of nodes $k = 32$, the same number of links $n = 80$, but strongly differing by their structure.

The first network has a completely symmetric structure. It is a five dimensional cube network H_5 with $2^5 = 32$ nodes. It is convenient to number the nodes by a five digit binary number, from $(0,0,0,0,0)$ to $(1,1,1,1,1)$. Each node is incident to five links connecting this node to five other nodes whose numbers differ by a single digit. So, for example, the node $(0,0,0,0,0)$ is incident to five links connecting it to the nodes $(1,0,0,0,0)$, $(0,1,0,0,0)$, $(0,0,1,0,0)$, $(0,0,0,1,0)$ and $(0,0,0,0,1)$. Cubic networks

I. Gertsbakh and Y. Shpungin, *Network Reliability and Resilience*,
SpringerBriefs in Electrical and Computer Engineering,
DOI: 10.1007/978-3-642-22374-7_2, © Ilya Gertsbakh 2011

have some optimal properties in the process of information delivery between nodes (see [18]) and because of that they are used as a frame for connecting computer stations (nodes) into a computer network.

The second network was created using the so-called preferential attachment method [13], Sect. 6.4. As suggested by Barabasi and Albert [2] in 1999, this method was supposed to reproduce the natural growth of large networks with a strongly nonuniform degree distribution and with the appearance of several nodes with very high number of adjacent links, called "hubs". Construction of a network starts with a "kernel" network $N_0 = (V_0, E_0)$ having a small number of nodes $|V_0| = m_0$ and several edges n_0, $|E_0| = n_0$. On each step of the construction, a new node v with d edges is added to the existing network, and the probability that v will be connected by an edge to an existing node w is proportional to the degree d_w of node w. We carried out this construction adding on each step a new node and five links, to obtain a network with 32 nodes and exactly 80 edges. This network—we call it *Prefnet*—has three nodes of degrees 13, 10, and 8, respectively, two nodes of degree 9, and the remaining nodes with degrees ranging from 2 to 6.

The third network is a modification of the network of terrorists responsible for the September 11 attack. Following the reference given in [3], p. 277, we took the description of this network from Valdis Krebs' website http://www.orgnet.com. This network has as its nodes the 19 hijackers who directly took part in the September 11 airplane attacks and 15 other people supporting them. The connections between all these people is represented by a network which has in total 34 nodes and 91 links. With regard to the node degree this network is highly non homogeneous. Its nodes can be classified into three groups. In the first, there are 6 nodes having small degree—one or two. There is a large "middle" group of 22 nodes, each one linked to 3–7 other nodes. The third group are the "hubs" linked to 8–16 other nodes, among them, there is one node linked to 16 nodes (representing the leader Mohammed Atta), one node linked to 14 nodes and a few nodes linked to 10–12 nodes. In order to carry out a valid comparison with H_5 and Prefnet, we modified slightly this network by deleting from it two nodes and 11 links. All connections of the hubs have been preserved. In further, we call this network *Ternet*. Comparing to Prefnet, Ternet is more irregular, its node degrees are more dispersed.

A.L. Barabasi mentioned the terrorist network in the context of the investigation of the behavior of scale-free networks [3, 19] under the intentional attacks on their most "important" nodes, which are their hubs. Networks with a single central node (so-called "spider networks") are very sensitive to the attack on their central node. Contrary to these networks, the scale-free networks without a spider are created in the process of self-organization and reveal a great degree of resilience and continue to function even after several of their hubs had been eliminated. So, A. Barabasi writes "... despite his central role, taking out Atta would not crippled the cell. The rest of the hubs would have kept the web together, possibly carrying out the attack without his help" (p. 223).

Ternet, Prefnet and H_5 have the same number of nodes (32) and the same average node degree 5. We decided to compare these three networks with respect to their resilience to the situation when they will be subject to a *random attack* on their nodes. These attacks mean that the nodes of each of these networks are subject to random independent failures taking place in random order. We remind that node failure means elimination of all links incident to this node.

We define two "degrees" of all-network damage (network failure):

Degree I. Network has disintegrated into separate isolated components of maximal size not exceeding max = 10. In the terms of a terrorist network, we assume that Degree I preserves considerable ability for hostile activities. For a network designed for maintaining normal life (food supply, emergency actions, defense, etc.) degree I means preserving relatively good capabilities to continue normal functioning for a relatively large part of network nodes.

Degree II. Network has disintegrated into separate isolated components of maximal size less or equal max = 3. Degree II means a very limited level of the potential for hostile actions for a terrorist network and very limited ability to maintain normal operation for any network providing supply/medical care, etc. Degree II means, in fact, a collapse of the network.

Formally, Degree I means that the maximal component has $3 < Size \leq 10$ and Degree II means that $Size \leq 3$.

Let us turn to the formal tools which will be used to describe the network gradual deterioration in the case of random node failures. Our D-spectrum technique is well-designed for this purpose. Let us remind that we consider a random node number permutation π, and turn *down* one node after another moving along π from left to right. In this process, the simulation algorithm follows up after the *size of the maximal connected component* in the network and remembers the positions of those nodes in the π whose failure signifies the entrance of the network into the states described above as Degree I and II of destruction. We will observe therefore a two-dimensional D-spectrum from which we will derive two marginal cumulative D-spectra denoted as $F(x; 1)$ and $F(x; 2)$, $x = 1, \ldots, 32$. We remind that $F(x; 1)$, the first cumulative D-spectrum, is the probability that the network is *DOWN* if $Y \leq x$ nodes have failed. For the first spectrum, *DOWN* means that the maximal connected component has the size *less or equal* 10. Similarly, the second spectrum $F(x; 2)$ is the probability that the maximal connected component in the network has the size less or equal 3 (Degree II). Figure 2.1 presents the graphs of the cumulative D-spectra for the Ternet, Prefnet and H_5. For Degrees I, II the maximal component size was max = 10, and max = 3, respectively.

All six curves fall apart into two distinct groups: the three left curves represent the first cumulative D-spectra for Ternet, Prefnet, and H_5, from left to right, respectively. Three curves on the right correspond, in the same order, to the Degree II of destruction. The most surprising fact is that Ternet is most vulnerable in both groups and H_5 has the highest resilience in both groups.

This fact contradicts the hypothesis that a scale-free self organized network is more resilient to the *random* attack on its nodes than a highly symmetric network. It turns out that the opposite is true, and more resilient is the symmetric network H_5. After a relatively large number of nodes fail (more than 22), the maximal component size becomes three (degree II), and all three networks behave similarly. The advantage of H_5 over Ternet and Prefnet becomes very small and vanishes with the

Fig. 2.1 The cumulative
D-spectra for three networks

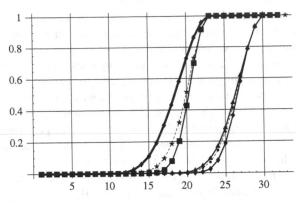

Fig. 2.2 Network failure
probability as a function of q

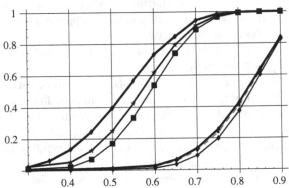

increase of x, which is quite natural: after most of nodes fail, H_5 has lost all its
initial symmetry and differs very little from any other network which had initially a
highly irregular degree distribution.

Note also that for the Ternet, for example, for $x = 25$ the probability of having
Degree II is near 0.3. With complementary probability about 0.7 the maximal connected component has the size greater than 3 but less or equal 32–25 = 7.

Next, let us compare the network failure probabilities $P(DOWN; q)$ for various
node failure probability values q. Figure 2.2 presents calculation results for three
networks Ternet, Prefnet and H_5, for two degrees of destruction. Degree I was
defined for max = 12 and Degree II for max = 3.

Six curves presented on Fig. 2.2 are naturally divided into two groups: the left for
degree I and the right for degree II. In each of these groups, the upper curve belongs
to Ternet, the lower—to H_5 and the middle—to Prefnet.

The difference between these curves is quite distinct in the left group. So, for
$q = 0.5$, H_5 is *DOWN* with probability near 0.18, while Ternet—with
probability 0.4. The difference between failure probabilities for H_5 and Ternet is
near 0.2 for $q \in [0.45, 65]$.

Table 2.1 *DOWN* probability as a function of node failure probability q

q	$P(DOWN; q)\ H_5$	$P(DOWN; q)$ Prefnet	$P(DOWN; q)$ Ternet
0.40	0.00227	0.053	0.134
0.45	0.070	0.126	0.245
0.50	**0.170**	0.250	**0.395**
0.55	0.332	0.424	0.567
0.60	0.538	0.619	0.733
0.65	0.740	0.793	0.846
0.70	0.889	0.914	0.947
0.75	0.967	0.975	0.986
0.80	0.994	0.996	0.998
0.85	0.9996	0.9999	1.0

As it is seen from Fig. 2.2, in the second group, all three networks behave similarly and the difference in their *DOWN* probability is less than 0.05 uniformly with respect to q values. This is explained by the fact that after a large number of nodes have already failed (on the average, more than $0.6 \times 32 = 19$) all networks look similar and the initial symmetry of H_5 has been strongly reduced by random node failures.

Table 2.1 presents the numerical data on *DOWN* probabilities for destruction degree I (max $= 12$).

So, for example, for $q = 0.5$, *DOWN* probabilities of H_5 and Ternet are 0.170 and 0.395, respectively, which means that H_5 is considerably more resilient.

2.2 Network Reliability Design

2.2.1 Introductory Remarks

Optimal network design is not a well-defined notion. It includes many particular problems, most of which are NP-complete and whose solution is a hard computation task.

The key words *Network Reliability Design* (NRD) produce on GOOGLE an astronomic number of references. Let us try to classify the main directions of the works in NRD.

One of the mathematically most interesting directions is the search of optimal network structures which produce maximal network reliability measure under given constraints on the number of network nodes and links, min cut set size or other combinatorial parameters, see e.g. [16, 17]. Close to this direction are the Ball-Provan bounds [1].

As a rule, in most applications, the researchers must compare and/or design a network system within a class of predetermined structures [7]. Usually the network "skeleton" is given, and the problem is the choice of links which must be introduced into the network or must be reinforced in the existing network to provide the desirable network reliability parameters. Typical for this line of research is the recent work [5]. It considers finding a minimum cost network with given reliability constraints. Mathematically, the problem is formulated in terms of integer programming, and powerful heuristics are suggested for its solution.

We must distinguish between two situations in the NRD. The first is when network reliability has a *known analytic* expression via the parameters of network node/link reliability, and the second is the case of unknown or practically non tractable expression of network reliability measure. The second situation is typical for networks of any realistic size, say having more than 10 nodes and 20 links.

In practice, the Monte Carlo methods remain the main tool for calculating real-size networks reliability for the purpose of NRD. One of the Monte Carlo reliability estimation advantages is that it can work with unreliable links and/or unreliable nodes, see e.g. [9], Chap. 8.

The NRD was investigated also using the *cross-entropy* modification of Monte Carlo [21], as well as the *genetic* Monte Carlo algorithms [14]. We believe that a combination of our methodology with genetic algorithms and the cross-entropy approach would produce good practical results.

Our approach for solving the NRD problem is based on using *reliability gradient* and so-called *BIM-spectra*, which are estimated by Monte Carlo simulation.

In this section we consider the following problem related to network design aimed at improving its reliability.

Suppose that we have a network with unreliable elements (nodes or edges). Network reliability is defined as the probability of its *UP* state, which is terminal connectivity. The following two operations are allowed:

1. Reinforcing a component, i.e. replacing it by a more reliable one. This operation can be applied to a given number of components.
2. Eliminating a component, i.e. a given number of components is supposed to be eliminated from the system.

The purpose of the above two operations is to achieve the maximal network reliability by "the best possible" choice of candidates for 1 and 2. The candidates can be any network edge in case of non reliable edges, or any node, which is not a terminal, in case of non reliable nodes. In practical terms, we must find the "most relevant" components for reinforcing and the "most irrelevant" ones for elimination.

2.2.2 The Dodecahedron Network

The network reliability design will be illustrated by an example of so-called dodecahedron network, see Fig. 2.3. It has 20 nodes and 30 edges, see [10]. Three nodes 1, 17 and 18 are terminals and the remaining 17 nodes are subject to failure. Node failure leaves the node itself intact, but all edges incident to this node are erased. Network *UP* state is defined as the terminal connectivity. Correspondingly, *DOWN*

Fig. 2.3 The dodecahedron network with three terminals (*bold*)

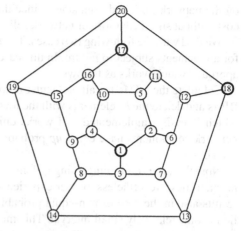

Table 2.2 Simulated BIM-spectra for nodes 2, 9, 20. M = 100,000 replications

x	j = 2	j = 9	j = 20
4	0.0011	0.0	0.0018
5	0.0074	0.0011	0.0091
6	0.025	0.0098	0.033
7	0.067	0.036	0.083
8	0.13	0.095	0.16
9	0.24	0.19	0.28
10	0.38	0.35	0.43
11	0.54	0.51	0.58
12	0.66	0.65	0.68
13	0.75	0.75	0.76
14	0.82	0.82	0.82
15	0.88	0.88	0.88
16	0.94	0.94	0.94
17	1	1	1

is a loss of terminal connectivity. For example, if nodes 20, 5, 6, 2, 13, 12, 11 fail, the network is in *DOWN* state.

As a preliminary step in NRD of this network, let us estimate the BIMs of network nodes. Analyzing the BIM-spectra for all 17 nonterminal nodes, we conclude that the node 20 is the most important and the node 2 is the second important. Table 2.2 presents the BIM-spectra for three nodes, and the domination of the node 20 over node 2 is clearly seen.

2.2.3 *Gradient-Based Network Reorganization*

Remind that by network reorganization we mean the following procedure. Suppose that k_1 network elements must be replaced by more reliable ones and that k_2 elements can be eliminated. The elimination of non important elements is a reasonable decision on the network project design stage, since this operation might reduce the network cost without significant loss of network reliability.

We deal with the following two cases. The first is the case of equal *up* probabilities for all elements subject to failure. In this case we suggest the BIM-spectra heuristic approach which works as follows.

Calculate the BIMs for all elements; reinforce the k_1 elements with the highest BIMs and delete the k_2 elements with the lowest BIMs. This is a heuristic procedure which is easily implemented and works quite well, especially for the case when network components have equal *up* probabilities or these probabilities are close to each other.

Note that after reinforcing a single element we are already in the case of non-equal probabilities. Nevertheless, by our experience, the BIM-spectra method gives good results also in the case of non-equal probabilities, when the values of probabilities lie in some relatively small interval. This may be explained by our observation that in most cases the "topological" or "structural" factor which is "concentrated" in the BIM (and is represented by the BIM-spectra) prevail over the numerical factor.

It is understood that elimination of a node means elimination of all edges incident to it. In case of unreliable edges, elimination of edges in the process of "trimming" always is made to avoid the appearance of isolated nodes.

Gradient based network reorganization heuristic

1. *Compute* element gradients for the initial network reliability by the Monte Carlo algorithm described in Sect. 1.8.3.
2. *Choose* the element having the maximal value of the product of the partial derivative times the element reliability increase, i.e. the *maximal* value of

$$\delta R_s = \delta p_s \cdot \frac{\partial R}{\partial p_s}, \quad s = 1, 2, \ldots, k$$

where $\delta p_s = p^\star - p_s$ and $p^\star$ is the reliability of the reinforced component. Denote by v the number of this element.
3. *Set* the reliability of element v, $p_v := p^\star$.
4. *Repeat* (1–3) k_1 times.
5. *Compute* element gradient vector for the network with the reinforced component reliability values.
6. *Choose* the element having the *minimal* value of the product

$$\delta R_s = p_s \cdot \frac{\partial R}{\partial p_s}.$$

Denote by w the number of this element and *Eliminate* the element w.
7. *Repeat* (5–6) k_2 times.

Let us illustrate how the above procedure works for the dodecahedron network.

Example 2.2.1 (*Unreliable nodes*).
We will consider the case of nonequal node reliability.

Table 2.3 Estimated reliability change for network with unreliable nodes, M = 100,000 replications

i	$q_i = 1 - p_i$	$\alpha = (\partial R/\partial p_i)(0.9 - p_i)$	$p_i(\partial R/\partial p_i)$
2	**0.3**	**0.0238**	0.083
3	0.25	0.02	0.099
4	0.3	0.022	0.077
5	0.3	0.0096	0.034
6	0.25	0.008	0.04
7	0.25	0.008	0.042
8	0.25	0.0069	0.0345
9	0.25	0.0063	0.031
10	0.30	0.0096	0.034
11	**0.25**	**0.0126**	0.063
12	0.25	0.0112	0.057
13	0.20	0.0074	0.059
14	0.20	0.0035	0.028
15	0.25	0.0054	0.027
16	0.25	0.012	0.059
19	0.20	0.0023	0.019
20	**0.2**	**0.0162**	0.13

The node *down* probabilities are given by the second column of Table 2.3. They range from 0.2 to 0.3. Suppose that we decide to reinforce *three* nodes and wish to eliminate *two* nodes. The reinforcing in our example means changing the chosen node by more reliable with the *up* probability $p^\star = 0.9$. The initial estimate of network reliability (based on $M = 10^5$ replications) equals $R_0 = 0.9076$.

The third column gives the values of the product defined in 2 of the above procedure. The values of the fourth column present the product defined in 6 of the same procedure. If component i is eliminated from the network, network reliability will *decrease* by the quantity shown in the fourth column.

We skip here the intermediate results of the calculations. The final results are the following. Nodes 2, 20, and 11 were chosen for the reinforcement. The nodes 15 and 19 are eliminated. The final reliability of the reorganized network is $R^\star = 0.9348$. Without elimination of nodes 15 and 19 the network reliability would have been slightly higher $R_1 = 0.9381$.

Remark In our example it is easy to determine intuitively the "most relevant" and "most irrelevant" nodes. Checking the third and the fourth column of Table 2.3 we see that nodes 2, 11 and 20 already have the required maximal values of α and nodes 15 and 19 have the minimal reliability decrease values. The impression is that it is enough to compute the gradient only once. For large networks with non-symmetrical and complex topology and with highly scattered component *up* probabilities, we must follow the gradient updating procedure as it has been described in the above heuristic.

The "reorganized" network is shown on Fig. 2.4. It is more reliable but it has lost its initial symmetry.

Fig. 2.4 Network with
unreliable nodes after
reorganization

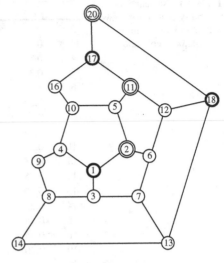

Fig. 2.5 Network with
unreliable edges

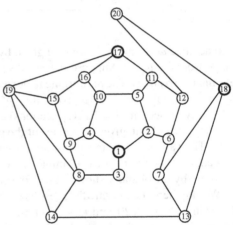

Example 2.2.2 (Unreliable edges).

Now let us consider slightly modified dodecahedron-type network presented on
Fig. 2.5. This network has reliable nodes and *unreliable edges*.

Suppose that the *up* probabilities of edges (1, 2), (7, 18), (11, 17) are 0.9. All
other edge *up* probabilities are 0.7. The initial estimate of network reliability (based
on 10^5 replications) equals $R_0 = \mathbf{0.7769}$.

Our purpose is to increase the network reliability at least up to $R^\star = 0.95$. Suppose
that we decided to reinforce the network by eliminating two (most irrelevant) edges
and by replacing the *required* number of edges by more reliable ones, with the *up*
probability equal 0.95.

Let us follow the above described gradient updating procedure. The initial gradient
values are presented in Table 2.4. From this table we see that the two candidates for

Table 2.4 Estimated Reliability gradient for network with unreliable edges

i	edge	p_i	$\partial R/\partial p_i$	i	edge	p_i	$\partial R/\partial p_i$
1	(1, 2)	0.9	0.1964	16	(8,14)	0.7	0.0574
2	(1, 3)	0.7	0.0653	17	(8,19)	0.7	0.0287
3	(1, 4)	0.7	0.0906	18	(9,15)	0.7	0.0219
4	(2, 5)	0.7	0.0836	19	(10,16)	0.7	0.0389
5	(2, 6)	0.7	0.1795	20	(11,12)	0.7	0.1435
6	(3, 8)	0.7	0.0642	21	(11,17)	0.9	0.1491
7	(4, 9)	0.7	0.0423	22	(12,20)	0.7	0.1051
8	(4.10)	0.7	0.0389	23	(13,14)	0.7	0.2078
9	(5, 10)	0.7	0.0334	24	(13,18)	0.7	0.0806
10	(5, 11)	0.7	0.0745	25	(14,19)	0.7	0.0632
11	(6, 7)	0.7	0.1894	26	(15,16)	0.7	0.0295
12	(6, 12)	0.7	0.0712	27	(15,19)	0.7	0.0230
13	(7, 13)	0.7	0.0515	28	(16,17)	0.7	0.0625
14	(7, 18)	0.9	0.1387	29	(17,19)	0.7	0.0954
15	(8, 9)	0.7	0.0253	30	(18,20)	0.7	0.1037

elimination are edges $18 = (9, 15)$ and $27 = (15, 19)$. Applying the gradient updating procedure (we skip the appropriate computations) we arrive at the following solution. Seven edges must be reinforced:

$$5 = (2, 6), 11 = (6, 7), 29 = (17, 19), 23 = (13, 14),$$
$$24 = (13, 18), 25 = (14, 19), 3 = (1, 4).$$

The reliability of the reinforced network is now $R^\star = 0.9547$.

Remark It is worth noting that if we take for reinforcing seven most relevant edges from Table 2.4 (i.e. the edges with largest $\partial R/\partial p_i$ values), we get a little smaller network reliability $R_1 = 0.9428$.

Example 2.2.3 ("Crossing" of two networks).

Consider the networks presented on Figs. 2.3 (call it network a) and 2.5 (call it network b). The components subject to failure in both networks are the *edges*. Let E_a and E_b be the edge sets of these networks.

In this example we demonstrate a procedure of "crossing" these two networks: we will obtain a new network with the 20 nodes and 30 edges, more reliable than a and b.

First, we will create a new network having the edge set which is a union of the edge sets of networks a and b; afterwards, we will reduce the united edge set by eliminating most irrelevant edges.

Suppose that all edges in both networks have *up* probability 0.7. Then the estimated values of the networks reliability are $R_a = 0.7778$ and $R_b = 0.6820$. Define network c with the same set of nodes and with the edge set $E_c = E_a \bigcup E_b$.

Fig. 2.6 Network with unreliable edges after "crossing"

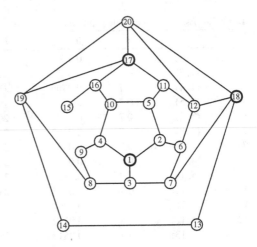

Network c has 20 nodes and 34 edges. Eliminate now the four *most irrelevant* edges by applying the gradient method.

Let us omit the intermediate calculations. As a result we have to eliminate the following four edges: (8, 14), (15, 16), (15, 19), (9, 15). The resulting network E_c is shown in Fig. 2.6.

The surprising fact is that this network reliability has increased to $R_c = 0.8375$. The way of "crossing" networks a and b to obtain a reliable network c is, in fact, an operation which might become a basis for a genetic algorithm for network reliability design.

2.3 Optimal Predisaster Design of Transportation Network

2.3.1 Formulation of the Problem. The Network

Optimal *predisaster design* means a decision on the network reinforcement policy in order to achieve the best reliability characteristics of the network under given budget constraint on the total cost of link reinforcement. In [24] the authors consider the road network reconstruction policy aimed at achieving higher reliability for the case when the traffic disruption is caused by road accidents. Interesting work [20] considers optimal road network design to minimize the damage from a future earthquake in the vicinity of Istanbul. The road network considered in [20] has 25 nodes and 30 edges, edges are subject to damage in the case of earthquake, and each edge e has a cost for its reinforcement $c(e)$. The reinforcement in the above cited work means that the initial edge failure probability $q(e)$ is being reduced to zero for the cost $c(e)$. The goal is to maximize the average probability of $s - t$-connection between five pairs of terminal nodes, subject to given budget B on the total cost of reinforcement works.

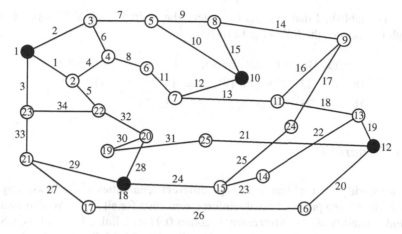

Fig. 2.7 Road network with 25 nodes, 34 links and 4 terminals (*bold*)

Gertsbakh and Shpungin [11, 25] consider the same road network and suggest a rather straightforward knapsack-type algorithm to maximize the minimal probability of the *s-t* connection. Interesting to note that the solution presented in [25] practically coincides with the solution of [20].

In this section we consider an optimal road network reinforcement policy on an example of a network with 25 nodes, 34 edges (links) and four terminals. The aim of this policy is to maximize the probability of terminal connectivity. The network is shown on Fig. 2.7.

Since there are many different ways to state and solve the problem, we will consider several scenarios for the initial data and the problem formulation.

2.3.2 Scenario A

All links (edges) have the same failure probability $q(e) = 0.3$ in the case of the earthquake, all link failures are independent events. We can choose any set of links for the reinforcement. If a link is reinforced, its failure probability reduces from $q(e) = 0.3$ to $q^*(e) = 0.1$. The reinforcement cost is the same for all links. The network has the initial probability of terminal connectivity $R_0 = 0.474$, as it can be established using Monte Carlo simulation. Our goal is to raise this probability to a relatively high level $R^* = 0.85$ by choosing the minimal number of links to be reinforced. (This formulation is dual to the situation considered in [20], where the reliability was maximized under cost constraints).

Since all links are equally reliable, the main tool for choosing the links for reinforcement is the analysis of their BIMs. Obviously, the best candidates will be the links with the highest BIM values.

It was established that in order to achieve the desired level of $R^\star = 0.846$ it is enough to reinforce the following 12 links:

$$2 = (1, 3), 3 = (1, 23), 7 = (3, 5), 12 = (7, 10), 14 = (8, 9),$$
$$18 = (11, 13), 24 = (15, 18), 29 = (18, 21), 30 = (19, 20),$$
$$31 = (19, 25), 32 = (20, 22), 33 = (21, 23).$$

2.3.3 Scenario B

Now we modify the problem and assign different reliabilities to the links, ranging from 0.6 to 0.8, but preserve equal reinforcement costs for all links. We also assume that link reliability after reinforcement becomes 0.9 for all links. The goal, as in Scenario A, remains the same—achieving the reliability $R^\star = 0.850$ by minimal cost. All necessary information for the solution is presented in Table 2.5. The third column shows the link reliabilities $p(e) = 1 - q(e)$. 11 links are assigned $p(e) = 0.6$, 11 links—$p(e) = 0.7$, and 12 links—$p(e) = 0.8$. The initial network reliability is $R_0 = 0.5011$. The fifth column gives the values of the partial derivatives, and the sixth—the reliability increase α after replacing a link having reliability p by a reinforced link with reliability $p^\star = 0.9$. As in the previous Scenario A, reinforcement of 12 links (marked by $\star$) achieve the goal and provide the desired level of reliability $R^\star = 0.850$. It is a remarkable fact that these links are located in "one run" from the sixth column of the table as the links with the maximal value of

$$\alpha = \frac{\partial R}{\partial p(e)} \cdot (0.9 - p(e)).$$

Theoretically, after choosing one or two links for the reinforcement, the link reliabilities change and the gradient vector changes too. Interesting to note that the initial ordering from column 6 remains the same. Thus, in Scenario B, the crucial role plays the gradient vector and the corresponding Monte Carlo procedure for BIMs estimation, see Sect. 1.8.

2.3.4 Scenario C

In this Scenario we introduce *different* reinforcement costs for different links, see the fourth column of Table 2.5. The links are divided into five categories, according to the costs of their reinforcement. The costs are nominated to be $c(e) = 1$–4 or 5. Low cost $c(e) = 1$ reflects minor reinforcement work needed to be carried out, the cost $c(e) = 5$ corresponds to fundamental reconstruction. The reinforcement goal remains the same: to achieve $R^\star = 0.850$ for the minimal cost. Note, that the total cost for the choice made in Scenario B (see the starred values in the Table 2.5, in column 5) is

Table 2.5 Edges, costs and the initial $\partial R/\partial p(e)$ values

i	edge e	$p(e)$	cost $c(e)$	$\partial R/\partial p(e)$	$\alpha = \partial R/\partial p(e) \cdot (0.9 - p(e))$	$\alpha/c(e)$
1	2	3	4	5	6	7
1	(1, 2)	0.8	1	0.091793	0.0092	0.0092
2	(1, 3)	0.6	2	0.137580	0.0413*	0.0206
3	(1, 23)	0.7	3	0.140512	0.0281*	0.0094
4	(2, 4)	0.8	4	0.091213	0.091	0.0023
5	(2, 22)	0.6	5	0.088629	0.0266*	0.053
6	(3, 4)	0.7	1	0.068969	0.0138	0.0138
7	(3, 5)	0.8	2	0.207425	0.0207	0.0104
8	(4, 6)	0.6	3	0.062008	0.0186	0.0062
9	(5, 8)	0.7	4	0.044431	0.0089	0.0022
10	(5, 10)	0.8	5	0.102330	0.00102	0.020
11	(6, 7)	0.6	1	0.062211	0.0187	0.0187
12	(7, 10)	0.7	2	0.129837	0.0260	0.013
13	(7, 11)	0.8	3	0.119049	0.0119	0.0040
14	(8, 9)	0.6	4	0.103155	0.0309*	0.0077
15	(8, 10)	0.8	5	0.052998	0.0106	0.0021
16	(9, 11)	0.8	1	0.062072	0.0062	0.0062
17	(9, 24)	0.6	2	0.094320	0.0283*	0.0142
18	(11, 13)	0.7	3	0.210631	0.0421*	0.0140
19	(12, 13)	0.8	4	0.272573	0.0273*	0.0068
20	(12, 16)	0.6	5	0.079393	0.0238	0.0048
21	(12, 25)	0.7	1	0.128011	0.0256	0.0256*
22	(13, 14)	0.8	2	0.076247	0.0076	0.0038
23	(14, 15)	0.6	3	0.105462	0.0316*	0.0105
24	(15, 18)	0.7	4	0.158136	0.0316*	0.0079
25	(15, 24)	0.8	5	0.070892	0.0071	0.0014
26	(16, 17)	0.6	1	0.077771	0.0233	0.0233*
27	(17, 21)	0.7	2	0.068477	0.0137	0.0068
28	(18, 20)	0.8	3	0.152753	0.0153	0.0051
29	(18, 21)	0.6	4	0.131870	0.0396*	0.0099
30	(19, 20)	0.7	5	0.131850	0.0264*	0.0053
31	(19, 25)	0.8	1	0.112046	0.0112	0.0112
32	(20, 22)	0.6	2	0.147710	0.0433*	0.0216*
33	(21, 23)	0.8	3	0.120779	0.0242	0.0080
34	(22, 23)	0.8	4	0.057959	0.0058	0.0014

$$\sum_{(e:2,3,5,14,17,18,19,23,24,29,30,32)} c(e) = 41.$$

Now the crucial role in the choice of links for reinforcement belongs to the ratio

$$\alpha/c(e),$$

which is the reliability increase per unit cost. Following the logic of the heuristic Knapsack solution, we find out a group of three links which have the *largest* values of this ratio. They are the links 21, 26, 32, see column 7 of Table 2.5. We raise their reliability to the value $p^\star = 0.9$ and the probability of terminal connectivity becomes equal $R_1 = 0.593$.

Now recompute the gradient vector and repeat the same procedure. The choice will be to reinforce links 2, 6 and 11, which gives $R_2 = 0.659$. (We omit the intermediate calculations).

The next iteration adds links 12, 18, 31 and the result is $R_3 = 0.738$.

The fourth iteration adds links 3, 7 and 17 and results in $R^\star = 0.795$.

Finally, only the fifth iteration which adds links 19, 27 and 29 achieves the desired reliability and even exceeds it: $\mathbf{R = 0.864}$. The total cost of this choice is

$$C = \sum_{(e:21,26,32,2,6,11,12,18,31,3,7,17,19,27,29)} c(e) = \mathbf{31}.$$

This is a considerable saving comparing to the cost of the result achieved in Scenario B. If 12 edges with *minimal* cost would have been chosen for the reinforcement, then the cost would be only 15, but the reliability would be very low—$R = 0.794$.

To underline the difficulty arising in finding the "optimal" solution, let us mention that we could lower the cost C by 4 units by excluding the link 19, but the reliability would become $\mathbf{0.842}$, which is below the desired level. Excluding link 29 would produce the same cost 31–4 = 27 but with reliability $\mathbf{0.833}$. In other words, there is a principal difficulty in comparing the "best" solution with the "second best" because of different costs.

The *sequential* application of the knapsack principle (i.e. choosing for reinforcement the link which on each step has the largest value of α among the non reinforced links) is, by our believe, a reliable way of getting good if not the best possible solution. In applying this principle we can manipulate by the size of the step, i.e. choose simultaneously several links for reinforcement. In the above example we took *three* links and the whole solution was carried out in five steps. We have tried several other strategies, and in particular, tried to choose a *single* link on each step, so that the whole solution was made in 14 steps with recomputing after each step the gradient vector.

Surprisingly, we obtained a slightly better result: reinforcing only 14 links (instead of 15) we achieved reliability $\mathbf{R^\star = 0.873}$ for the same cost of $C = 31$. The links to reinforce are the following:

$$21, 32, 2, 26, 11, 31, 12, 18, 17, 3, 7, 27, 29, 20.$$

It should be noted that the choice of the links to be reinforced on steps 9–14 was based on very small differences between the values of $\alpha/c(e)$ which might have appeared as a result of a statistical error in Monte Carlo simulation of the gradient. Since the difference $\mathbf{R^\star - R} = 0.009$ exceeds the statistical error caused by reliability estimation for $M = 100,000$ replications, we are inclined to assume that the single-step strategy is preferable.

2.4 Network Disintegration

2.4.1 Introduction

In this section we consider the following situation. Network components (links) start to fail, in random order, one after another. In reality, this may happen as a result of a sequence of heavy road accidents, natural disasters (floods, earthquakes, fires) or an intentional "enemy attack". Sooner or later, one or several terminals gets isolated from other terminals, and the process continues until the network completely collapses. Network collapse, by our definition, is network disintegration into isolated clusters, i.e. the network state in which each terminal becomes isolated from all other terminals.

Our goal is to provide a probabilistic description of this process. In particular, we are interested in the probability $P(x; W)$ that after a failure of x links the networks has disintegrated into W clusters.

Several examples of this kind are presented in [15] and [12]. The main formal tool for investigating network disintegration is its marginal D-spectra, see Sect. 1.3. The network disintegration we will illustrate by the example of the transportation network considered in the previous section. This network has 25 nodes, 4 of which are terminals, and 34 links. Its *UP* state corresponds to $W = 1$, i.e. when there is only one cluster (all four terminals are connected to each other). In the process of gradual link failures, the network enters state $W = 2$, $W = 3$ and eventually disintegrates into four isolated clusters ($W = 4$). The process of gradual link failures can be considered in various time frames. For example, a reasonable assumption is that links failures take place according to an "external" Poisson shock process with parameter λ, see e.g. [4, 8]. Another assumption is to imagine that links have random i.i.d. lifetimes $\tau_1, \ldots, \tau_{34}$ with a continuous CDF $H(t) = P(\tau_i \leq t)$, see [22, 23]. Important is the fact that links fail in *random* order, and all possible orders are equally probable. This is automatically guaranteed by the i.i.d. assumption. We will not specify the temporal process of link failures appearance and will count the "time" in the units of failed links. This is equivalent to the assumption that links fail at the instants $t = 1, 2, 3, \ldots$

2.4.2 Discrete Densities of the Transition Times

The discrete marginal densities for the location of the first, second and third anchor are presented in Table 2.6, columns 2–4. We see that the appearance of two clusters happens in the interval $x \in [3, 23]$. Elementary calculations give that the average number of link failures for this event is $m_1 = \sum_{x=1}^{34} x \cdot f^{(1)}(x) = 10.4$. The appearance of three clusters happens in diapason $x \in [6, 27]$, and the average is $m_2 = 14.3$. The appearance of four clusters is described by the third spectrum (the last column) and takes place for $x \in [8, 32]$, with the average $m_3 = 18.8$. Summing up, two clusters appear on the average after $x = 10$ links fail, and each next cluster appears, on the

Table 2.6 Edge spectra for the transport network

x	$\mathbf{f}^{(1)}$	$\mathbf{f}^{(2)}$	$\mathbf{f}^{(3)}$
1	0	0	0
2	0	0	0
3	0.00216	0	0
4	0.00775	0	0
5	0.01799	0	0
6	0.03674	0.00014	0
7	0.06284	0.00127	0
8	0.09842	0.00593	0.00001
9	0.13231	0.01729	0.00015
10	0.15362	0.04225	0.00108
11	0.15198	0.08104	0.00488
12	0.12819	0.12221	0.01421
13	0.09072	0.15000	0.03248
14	0.05604	0.14950	0.05813
15	0.03230	0.12982	0.08072
16	0.01663	0.10444	0.09758
17	0.00740	0.07595	0.10963
18	0.00307	0.05026	0.10699
19	0.00118	0.03204	0.10134
20	0.00050	0.01846	0.09017
21	0.00012	0.00982	0.07558
22	0.00003	0.00520	0.06189
23	0.00001	0.00258	0.04916
24	0	0.00107	0.03741
25	0	0.00043	0.02765
26	0	0.00020	0.01923
27	0	0.00010	0.01336
28	0	0	0.00877
29	0	0	0.00536
30	0	0	0.00248
31	0	0	0.00137
32	0	0	0.00037
33	0	0	0
34	0	0	0

average, after 4–5 link failure. The graphs of the discrete densities $\mathbf{f}^{(r)}$ are presented in Fig. 2.8.

2.4.3 The Cumulative D-Spectra and State Probabilities

The cumulative D-spectra $F^{(i)}(x)$, $i = 1$–3 are presented in Table 2.7 in the columns 2–4. We remind that $F^{(1)}(x)$ is the probability that after x links have failed, the network has two or more clusters. Similarly, $F^{(2)}(x)$ is the probability that after x

Fig. 2.8 Discrete marginal densities of cluster appearance

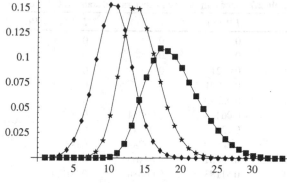

Fig. 2.9 Cumulative D-spectra $F^{(1)}(x)$, $F^{(2)}(x)$, $F^{(3)}(x)$

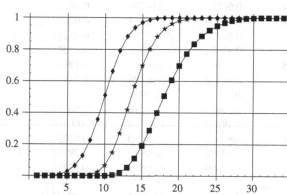

link have failed, there are at least three clusters. Finally, $F^{(3)}(x) = P(x, W = 4)$ is the probability that the network has four clusters. The last two columns give the probabilities that there are exactly 2 or 3 clusters, respectively. Obviously,

$$P(x; 2) = F^{(1)}(x) - F^{(2)}(x), \quad P(x; 3) = F^{(2)}(x) - F^{(3)}(x),$$

see Sect. 1.3.

Figure 2.9 presents the graphs of all three cumulative spectra.

2.4.4 Network State Probabilities as Function of q

In conclusion, we consider the network state probabilities as a function of link failure probability q. According to Sect. 1.3,

$$P(UP) = P(1 \ cluster; q) = 1 - \sum_{x=1}^{34} F^{(1)}(x) q^x (1-q)^{(34-x)} 34!/((x!(34-x)!),$$

Table 2.7 Edge cumulative D-spectra and state probabilities, $M = 10^5$

x	$F^{(1)}(x)$	$F^{(2)}(x)$	$F^{(3)}(x) = P(x;4)$	$P(x;2)$	$P(x;3)$
1	0	0	0	0	0
2	0	0	0	0	0
3	0.00216	0	0	0.00216	0
4	0.00991	0	0	0.00991	0
5	0.02790	0	0	0.02790	0
6	0.06464	0.00014	0	0.06450	0.00014
7	0.12748	0.00141	0	0.12607	0.00141
8	0.22590	0.00734	0.00001	0.15250	0.00733
9	0.35821	0.02463	0.00016	0.33358	0.02447
10	0.51183	0.06688	0.00124	0.44495	0.06564
11	0.66381	0.14792	0.00612	0.51589	0.1418
12	0.79200	0.27013	0.02033	0.52187	0.2498
13	0.88272	0.42013	0.05281	0.44259	0.36732
14	0.93876	0.56963	0.11094	0.36913	0.45869
15	0.97106	0.69945	0.19166	0.27161	0.50779
16	0.98769	0.80389	0.28924	0.18380	0.51465
17	0.99509	0.87984	0.39887	0.11525	0.48097
18	0.99816	0.93010	0.50586	0.06806	0.42424
19	0.99934	0.96214	0.60720	0.03720	0.35494
20	0.99984	0.98060	0.69737	0.01924	0.28323
21	0.99996	0.99042	0.77295	0.00954	0.28323
22	0.99999	0.99562	0.83484	0.00437	0.21747
23	1	0.99820	0.88840	0.00180	0.16078
24	1	0.99927	0.92141	0.00073	0.11420
25	1	0.99970	0.94906	0.00030	0.07786
26	1	0.99990	0.96829	0.00010	0.05064
27	1	1	0.98165	0	0.03161
28	1	1	0.99042	0	0.01835
29	1	1	0.99578	0	0.00958
30	1	1	0.99826	0	0.00422
31	1	1	0.99963	0	0.00174
32	1	1	1	0	0
33	1	1	1	0	0
34	1	1	1	0	0

$$P(2;q) = P(2\ clusters;q) = \sum_{x=1}^{34} (F^{(1)}(x) - F^{(2)}(x))q^x(1-q)^{(34-x)}34!/((x!(34-x)!),$$

$$P(3;q) = P(3\ clusters;q) = \sum_{x=1}^{34} (F^{(2)}(x) - F^{(3)}(x))q^x(1-q)^{(34-x)}34!/((x!(34-x)!),$$

$$P(4;q) = P(4\ clusters;q) = \sum_{x=1}^{34} F^{(3)}(x)q^x(1-q)^{(34-x)}34!/((x!(34-x)!).$$

Table 2.8 Network state probabilities

q	$P(UP)$	$P(2)$	$P(3)$	$P(4)$
0.1	0.981	0.019	0.0003	0.000001
0.2	**0.817**	**0.160**	**0.021**	**0.016**
0.3	0.475	0.351	0.145	0.029
0.4	0.148	0.336	0.337	0.155
0.5	0.037	0.168	0.389	0.406

Table 2.8 presents the numerical values of these probabilities for $q = 0.1(0.1)0.5$. We see from it how the probabilities are distributed between the states of the network. For example, for $q = 0.2$ the network is UP with probability 0.817, and the most of the remaining probability belongs to the state with exactly two clusters. For $q = 0.4$, the network is UP with probability 0.148 only, and with probabilities about 1/3 there are two or three clusters, respectively.

2.4.5 Network Resilience

We remind that the probabilistic resilience of the network $\mathrm{res}_{pr}(\mathbf{N}; \beta)$ is the largest value x such that $1 - F(x) > 1 - \beta$, where $F(x)$ is the cumulative D-spectrum, see Sect. 1.3.2. Assume $\beta = 0.075$. From Table 2.7, column 2, we see that with respect to network terminal connectivity,

$$\mathrm{res}_{pr}(\mathbf{N}; \beta) = 6,$$

which gives a rather low value of relative resilience ([6], p. 435)

$$\eta = \frac{\mathrm{res}_{pr}(\mathbf{N}; \beta)}{34} = 0.17.$$

If the network failure is defined as the total collapse (4 clusters), then from the fourth column of Table 2.7. we see that

$$\mathrm{res}_{pr}(\mathbf{N}; \beta) = 13,$$

which gives the relative resilience $\eta = 13/34 = 0.38$.

These η values reflect rather low network reliability which is quite understandable because the network is not dense and the average node degree $\overline{d} = 34 \times 2/25 = 2.72$.

References

1. Ball, M.O., Colbourn, C.J., and J.S. Provan. 1995. Network reliability. *Handbook of OR and MS Chapter 11*. New York: Elsevier.
2. Barabasi, A., and R. Albert. 1999. Emergence of scaling in random networks. *Science* 286(5439): 509–512.

3. Barabasi, Lazslo Albert. 2003. *Linked*. Penguin Group (USA) Inc.
4. Barlow, R.E., and F. Proschan. 1975. *Statistical theory of reliability and life testing.* New York: Holt, Rinehart and Winston, Inc.
5. Beraldi, P., Bruni, M., and F. Guerriero. 2010. Network reliability design via joint probabilistic constraints. *IMA Journal of Management Mathematics* 21(2): 213–226.
6. Brandes, U., and T. Erlebach eds. 2005. *Network analysis-methodological foundations.* Heidelberg: Springer-Verlag Berlin.
7. Dasgupta, M., and G.P. Biswas. 2010. Reliability measurement and enhancement of the communication network. *International Journal of Computer Applications* 1(9): 11–18.
8. Gertsbakh, I.B. 1989. *Statistical reliability theory*. New York: Marcel Dekker, Inc.
9. Gertsbakh, Ilya and Yoseph Shpungin. 2009. *Models of network reliability: analysis, combinatorics, and Monte Carlo*. CRC Press.
10. Gertsbakh, I., and Y. Shpungin. 2010. Network reliability design: combinatorial and Monte Carlo approach. *Proceedings of the International Conference on Modeling and Simulation (MC-2010)*, 88–94. Banff, Canada.
11. Gertsbakh, I., and Y. Shpungin. 2010. Optimal Reliability Design of Transportation Network. *Proceedings of the 10th International Conference on Reliability and Statistics in Transportation and Communication - 2010*. Riga, Latvia 20–24.
12. Gertsbakh, I., and Y. Shpungin. 2012. Stochastic Models of Network Survivability, *Quality Technology and Quantitative Management*, Special Issue devoted to S. Zacks 9(1) (to appear).
13. Kolaczyc, Erik. 2009. *Statistical analysis of network data*, Springer Science+Busineee Media.
14. Kumar, P., Parida, P., and M. Gupta. 2002. Topological design of communication network using multiobjective genetic optimization. *Evolutionary Computation* 1: 425–430.
15. Levitin, G., Gertsbakh, I., and Y. Shpungin. 2010. Evaluating the damage associated with intentional network disintegration. *Reliability Engineering and System Safety* 96(4): 433–439.
16. Lomonosov, M.V., and V.P. Polesskii. 1971. An upper bound for the reliability of information network. *Problems of Information Transmission* 7: 337–339.
17. Lomonosov, M.V., and V.P. Polesskii. 1972. Lower bound on network reliability. *Problems of Information Transmission* 8: 45–53.
18. Mitzenmacher, M., and E. Upfal. 2005. *Probability and computing*. Cambridge: Cambridge University Press.
19. Newman, M.E.J. 2010. *Networks: an introduction*. Oxford: Oxford University Press.
20. Peeta, S., Salman, S.F., Gunnec, D., and V. Kannan. 2010. Predisaster investment decisions for strengthening a highway network. *Computers and Operations Research* 37: 1708–1719.
21. Rubinstein, R.Y., and D.P. Kroese. 2007. *Simulation and Monte Carlo methods*, 2nd ed. Wiley.
22. Samaniego, F.J. 1985. On closure of the IFR under formation of coherent systems. *IEEE Transactions Reliability* 34: 69–72.
23. Samaniego, F.J. 2007. *System signatures and their applications in engineering reliability.* Berlin: Springer-New York.
24. Sanchez-Silva, M., Daniels, M., Lleras, G., and D.A. Patino. 2005. Transport reliability model for efficient assignment of resources. *Transportation Research, Part B* 39: 47–63.
25. Shpungin, Y., and I. Gertsbakh. 2010. Predisaster design of transportation network. *Transport and Telecommunication Scientific and Research Journal* 12(1): 4–11.

Index

I. Gertsbakh and Y. Shpungin, *Network Reliability and Resilience*,
SpringerBriefs in Electrical and Computer Engineering,
DOI: 10.1007/978-3-642-22374-7, © Ilya Gertsbakh 2011